TELEVISION IN THE EIGHTIES
THE TOTAL EQUATION

TELEVISION IN THE EIGHTIES
THE TOTAL EQUATION

Rex Moorfoot

Research
Dinah Garrett

Graphics
Martin Bronkhorst

BRITISH BROADCASTING CORPORATION

PICTURE CREDITS
APPLE COMPUTER (UK) LTD classroom computer, page 88; ATARI INTERNATIONAL (UK) INC. video computer system, page 27 (left); BBC Ceefax text, page 29 (left), Carfax receiver, page 51, radio-data receiver, page 52; BBC PHOTOGRAPH LIBRARY studio concert, page 9, studio play, page 10, schools broadcast, page 88; BRITISH TELECOM Goonhilly, page 15; GENERAL ELECTRIC large-screen projection, page 31; IPC video catalogue, page 80; JVC (UK) LTD video-cassette recorder, page 22 (top), video-disc player, page 24; ORACLE TELETEXT LTD Oracle text, page 29; PHILIPS ELECTRONICS video-cassette recorder, page 22 (bottom), Laservision, page 25 (left); POST OFFICE satellite, page 16, Prestel, page 28 (both), Confravision, page 97; RCA INTERNATIONAL LTD video-disc player, page 25 (right); SELECTV decoder box, page 21; SINCLAIR RESEARCH LTD personal computer, page 30; SONY (UK) LTD video-cassette recorder, page 23; TIMES NEWSPAPERS LTD video cassette, page 75; THORN EMI cassettes & records, page 57.

Published by the British Broadcasting Corporation
35 Marylebone High Street, London W1M 4AA

ISBN 0 563 20017 0 (hardback)
     0 563 20144 4 (paperback)
First published 1982
© Rex Moorfoot 1982

Set in 11/13pt Baskerville and printed in England by
Spottiswoode Ballantyne Limited, Colchester and London

CONTENTS

	ACKNOWLEDGEMENTS	7
	INTRODUCTION	8
1	THE TELEVISION SET	12
	Television systems:	13
	Television broadcasting; Satellite broadcasting; Cable television; Pay-tv; Home video; Video-cassette machines; Video-disc players; Cassettes versus discs	
	The Future Television Set:	26
	Video games; Home computers; Viewdata; Teletext; Home printers; Big screens; Stereo sound	
	The television home terminal	32
2	THE NEW TELEVISION MARKETS	34
	Consumer costs; Degree of choice; Audience size; Market value	
3	GROWTH OF THE MARKETS	37
	Number of television homes; Number of homes using Satellites, Cable, Pay-tv, Home video; Income forecasts; Programme demand in the United States and in the United Kingdom	
4	ANALYSIS OF THE NEW MARKETS	43
5	NEW SOURCES OF PROGRAMMES AND SERVICES	50
	Radio; Records; Cinema; Theatre; Sport; Newspapers; Magazines; Books; Education; Religion; Industry and Commerce	
6	THE CHALLENGE TO BROADCASTING	102
7	MEANS OF FUNDING	106
8	TECHNICAL QUESTIONS	110
9	LAWS AND AGREEMENTS	113
10	GOVERNMENT REGULATION	117
11	THE TOTAL EQUATION	123

ACKNOWLEDGEMENTS

I wish to record my appreciation of help given directly and indirectly by colleagues working in television in the United States, the United Kingdom and in Western Europe.

I wish also to thank Michael Tree who fostered the original idea; Tony Kingsford for the painstaking and constructive way he has edited the book; Linda Blakemore for the skilful design; Jennifer Fry for the photographs; and Audrey Wiener for coping with the typing.

Of my two closest colleagues, I am especially grateful to Martin Bronkhorst for his graphics illustration and for his technical advice on book production. I have to thank Dinah Garrett for the breadth of her research which made new observations possible. I am also very grateful to her for helping clarify the more obscure parts of the book.

My family and friends have shown much forbearance, in particular Ronald Blakey and Derek Stopps.

For the record, the idea for the book had first light at Bertorelli's Restaurant in Charlotte Street in London, and many subsequent discussions took place there.

INTRODUCTION

> **'QUIET,' SAYS ANNOUNCER**
> **B.B.C. Listeners Hear**
> Radio listeners who tuned in to the National programme at eight o'clock last night to hear a burlesque entitled "All at Sea," heard the announcer say, with a pause between each phrase, "Quiet, please, Quiet, please".

Daily Express,
17 August 1937.

I read that over breakfast the day after I joined the BBC. The scene is still very clear in my mind. 'All At Sea' was being produced by Martyn C. Webster in Studio 4 at the BBC in Birmingham. My job was to press a switch when the studio was ready to go on the air. The big wall clock ticked towards eight o'clock. The orchestra was tuning up. The red transmission light below the clock started to flash. Reginald Burstall, the conductor, tapped his baton to call the studio to order. I pressed the switch. Nothing happened, so I pressed it again and totally confused the situation. The announcer was caught unawares and his warning words were broadcast live to the nation.

Almost all broadcasting was live; everything happened in real time. For the performers, eight o'clock in Studio 4 was as significant as the curtain going up on a theatre first night. When the script called for a change in 'acoustic' scenery, say, from a rowdy boat deck to the captain's cabin, the actors concerned had to scurry from Studio 4, down the corridor to the heavily padded Studio 3, their temporary absence covered by a few bars of music. When it came to the repeat of 'All At Sea', the entire company had to reassemble in Studio 4 for the second 'live' performance at nine o'clock on the BBC Regional Programme the following night.

A 'live' performance in front of the microphone: Jelly d'Aranyi with Myra Hess at the piano in 1928 for a BBC radio programme from Savoy Hill.

Recording changed all that. With a recorded broadcast the time and place of producing the programme becomes totally dissociated from the time and place of its transmission. News and events recorded in the daytime can be broadcast in the evening. At first the means were cumbersome: the Blattnerphone, for example, consisted of huge spools of steel tape over a mile in length wound under stress which, if released, so it was said, could enmesh a man in its coils. When discs were first introduced, the coating was so soft that second copies had to be marked up for transmission to replace the worn-out rehearsal discs. Editing unwanted material was possible, but primitive. The needle was physically lifted by the thumb and first finger from one chinagraph mark to the next. Discs later gave way to magnetic tape, and editing became a matter of snipping out unwanted material with a pair of scissors, or copying one tape to another. Production techniques expanded and scheduling became simpler and more flexible.

The same story can be told for television. 'Live' television plays, for example, had the lack of mobility of theatre plays where the action is generally not seen and the drama lies in the reactions of the characters on stage. With recording, the producer gains the freedom to move his cast from location to location as in film-making. So the television viewer witnesses the action as it is happening, the hero in danger, say, as the ice breaks on the river, followed immediately by the same character returned home and dressed differently. It is no longer necessary to write padding scenes to cover the time for the actor to change his clothes or for his hair to dry!

When a programme is recorded, the actual broadcast can be at any time or on any day. This enables the controller of a television network to build an ideal programme mix, like a theatre playbill, which will provide a continuing experience for viewers of three to four hours. Popular programmes early on build up the audience, followed by programmes with strong narrative elements to 'inherit' the audience – a Western movie or a popular family series. New programmes are launched by being 'hammocked' between strong pulling programmes. The network controller has the flexibility to alter the programme mix on the night like a newspaper editor responding to news and events.

When a programme is recorded, it also becomes a physical product. It exists in its own right. It can be repeated. It can be leased or sold like films to other broadcasters, to cable and to satellite systems. As a physical product it can be sold over the counter to the public like a book or a record, in the form of video cassettes and video discs. This is the point at which this book takes up the story, as Television entered the Eighties.

I begin by assessing the new systems for distributing and paying for programmes – satellites, cable, video cassettes and discs, and pay-tv – and comparing them with television broadcasting. New sources of television programmes and services are explored – Radio, Records, Cinema, Theatre, Sport, Newspapers, Magazines, Books, Education, Religion,

The Importance of Being Ernest – an early television performance broadcast 'live' by the BBC from Alexandra Palace. Like the theatre, the action had to take place in one studio location.

and Industry and Commerce. I then spell out the challenges to television broadcasting from the new systems and from the potential sources of programmes and services, in terms of competition for audiences, programmes, talent and money. In the last chapter, The Total Equation, I consider the future role of broadcasting alongside the new systems, and what needs to be done to preserve the high standards and traditions of the best of broadcasting in the United Kingdom and Europe in the face of competition.

Many examples are quoted of developments in America which is generally ahead of the world in the exploitation of the new systems. For convenience, I have taken $2 to the £1 and worked on 1980/81 prices. Forecasts and estimates have been drawn from a variety of sources to give a consensus view and to provide a basis for comparison. The aim has been to simplify, to round up figures, and not to lose sight of the underlying principles in a mass of detail. It is a task made more difficult by the increasing pace of events.

Everything in the book is here with us in 1982, not on the drawing board, or in pilot production. What I have tried to do is to set down the total equation of Television in the Eighties, albeit with the eyeline of a lifetime in broadcasting. The views are my own.

Rex Moorfoot, Teddington
16 August 1982

1 THE TELEVISION SET

Since the first transmission in 1936 in Britain, the television set has done little more than receive broadcast programmes. The television set is capable, however, of receiving programmes from many other sources, via cable and satellite on a pay-as-you-view basis if required. Programmes can also be played into the television set from video cassettes and video discs. Other facilities can be plugged in to provide additional services: video games, home computers, the telephone to display information on the screen, and video printers to supply hard copy. This chapter describes the nature and growth of the new sources of programmes and the additional services, and how they will affect the design of the television set of the future.

There are over twenty million television sets in the United Kingdom. A good colour set sells at around £500, so this represents a capital investment by viewers in the order of £1 billion. Such an investment could well inhibit the introduction of new services unless existing equipment can be utilised. In Britain, however, two-thirds of all television sets are rented for about £2 a week, and, significantly, the practice is being extended to the hiring of other equipment, for example, video cassette players. This is going to be a unique factor in the development of television in Britain.

Price and rental charges vary according to screen size and special features like remote control. For £500 the screen probably measures 27″ diagonally, with an aspect ratio of 4:3 in the tradition of the cinema. A wider screen will soon be possible following the pattern of the modern cinema. This will improve the television presentation of some events: in horse racing, for example, the whole field will be seen in greater detail. It will also be technically possible in the near future to have an absolutely flat screen of improved luminosity.

Definition depends on the number of electronic lines that make up the picture. John Logie Baird, one of the acknowledged inventors of television, made use of twenty-five lines. When television started officially in the United Kingdom in 1936, the BBC made use of 405 lines. All European broadcasters, including the BBC, now operate on 625 lines. American television uses 525 lines and is based on the NTSC colour system as opposed to the French system SECAM and the Anglo-German system, PAL. The adoption of the SECAM and PAL systems throughout the world has more or less followed the historic colonial pattern of the French and the British Empires, except for Russia which adopted

SECAM, whereas the NTSC system has broadly been adopted by countries in the American sphere of influence, Japan and the Philippines in the Far East, for example, and South America and Canada.

So the television services throughout the world have not only different line standards, but also incompatible colour systems. The only way programmes from one service can be distributed to another is to pass the signal through a computer-based 'standards converter'. Further complications will arise as services based on 1000 lines or more are developed to improve definition on large screens, matching the quality of cinema presentation.

TELEVISION BROADCASTING

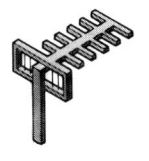

An aerial is the link between the television set and the television transmitter. Television needs 1000 television transmitters to cover the United Kingdom. Radio requires only one powerful low-frequency transmitter. This is because radio frequencies do not attenuate rapidly and some are reflected by atmospheric layers in a series of hops round the earth's surface. To carry pictures and sound, television has to use much higher frequencies which travel only as far as the eye can see, like a beam of light. A television transmitter, even on a hilltop, has a range of only about forty-five miles. Nonetheless, transmission can be omnidirectional; thus the transmitter at the Crystal Palace outside London reaches five million homes.

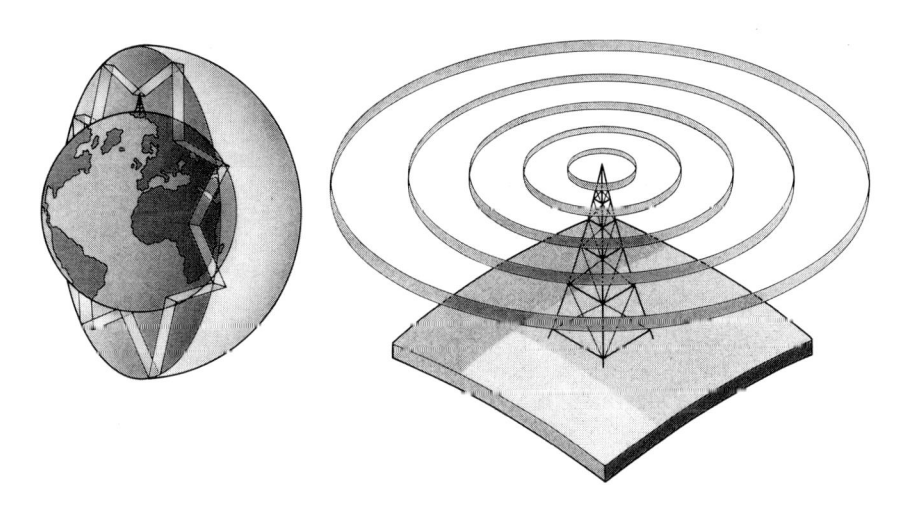

The television signal carries pictures and sound, and the sound can be in stereo or two languages. There is also sufficient 'space' to carry additional information in the form of hundreds of captions, which the viewer can select at random. This facility has been developed in the United Kingdom as a service of news, reference material and sub-titles, called Teletext.

The range of a transmitter is determined by its power. If the power is reduced, the signal falls off. The BBC transmitter at Crystal Palace operates on 1000 kilowatts. In the United States some transmitters operate on as little as 10 watts, just enough energy to operate a night-light lamp. These so-called Multi-Distribution Systems (MDS) were envisaged originally for the dissemination of local data transmission. They were first exploited for television to link apartment houses in Sacramento, California. The biggest installation is now in Anchorage, Alaska, where one transmitter serves 10,500 individual houses.

The way television broadcasting is paid for and the services provided vary from country to country. In Western European countries viewers usually pay a licence fee. In the United Kingdom it costs about £1 a week per household for a licence, and this pays for the two BBC colour networks, BBC-1 and BBC-2 and BBC Radio. The ITV commercial services are funded by advertising. For a total cost of £1200 million per year both organisations provide a variety of programmes, averaging forty hours a day. Each network has a teletext service, CEEFAX on BBC-1 and BBC-2 and ORACLE on ITV. Channel 4, the second commercial network, is scheduled to begin in 1982, and it is planned to introduce breakfast television in 1983.

In the United States there are 1000 television stations across the country: 737 are commercial stations funded in the main by advertising, and 263 are non-commercial and educational stations funded largely by private and public subscriptions. A typical city has six different stations: three affiliated with the national networks, ABC, CBS and NBC; two independent commercial stations; and one non-commercial station, usually associated with the Public Broadcasting Service – the PBS network. Together they provide 120 hours of programmes a day. The total annual cost of commercial television in America is the equivalent of about £2500 million, which works out at £32 per television household a year. And the average household has the set switched on for nearly seven hours a day, approaching double the average for the United Kingdom.

SATELLITE BROADCASTING

A satellite extends the range of television. A television signal can now reach over a country the area of the United States or India, or span the world, by bouncing off satellites, rather like the natural phenomena of radio signals. Only three satellites are required to span the world. A communications satellite carries up to twenty-four 'transponders' to receive television signals and to retransmit them. Very much higher frequencies have to be used than for terrestrial transmission and the distances are infinitely greater. To receive transmissions from space parabolic dish aerials are required to focus on the very weak signals.

Telstar in 1962 was the first satellite to relay television. As it circled

THE TELEVISION SET

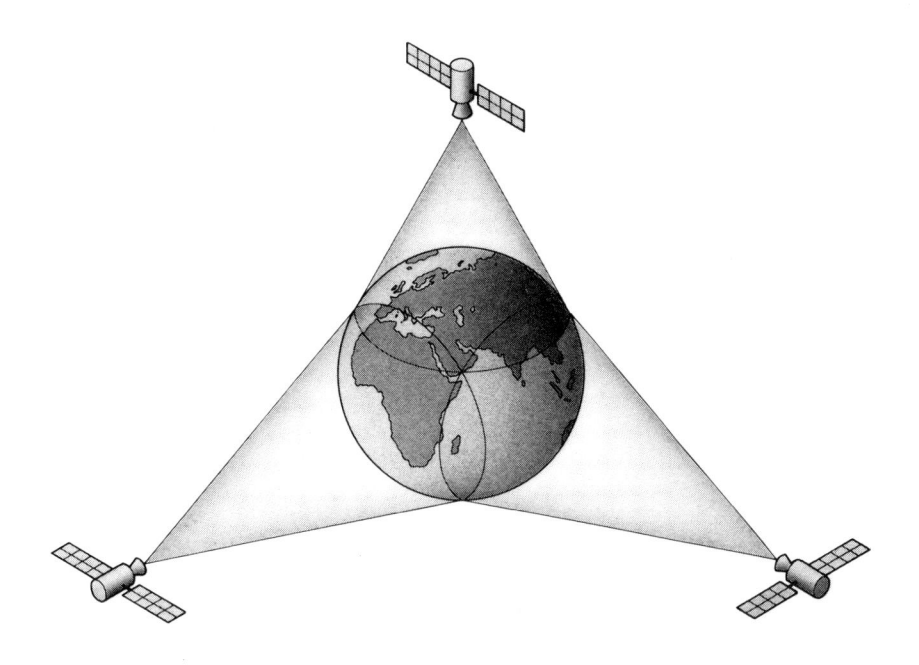

A dish aerial at Goonhilly, one of the first satellite tracking stations.

the earth, giant dish aerials, like Goonhilly in south-west England, tracked its twenty-minute journey from horizon to horizon, high enough to encompass London and New York in one arc and to make possible intercontinental television broadcasting for the first time. One senior American television executive maintains that a brass plaque should be mounted on the door of Room 6025 at the BBC Television Centre to commemorate the arguments, recriminations and agreements reached between the British and American networks in the early sixties over the allocation of time and order of appearance in the twenty-minute period.

Such problems no longer exist. Present-day geo-stationary satellites encircle the earth at 22,300 miles on a critical equatorial orbit which allows the satellite to rotate synchronously with the earth and thus appear stationary at all times.

In 1977 the World Administrative Radio Conference allocated five satellite channels to every country in the world. The Conference also defined the areas of reception, the national footprints as they are called. By their nature these will spread over neighbouring countries, creating new problems in international broadcasting.

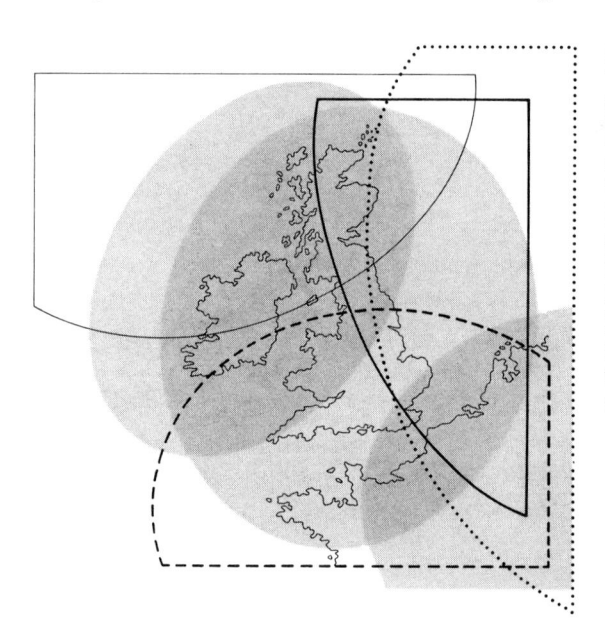

National footprints. The large shaded area is the footprint for the United Kingdom. The Irish footprint spreads over Scotland and Wales; Iceland over N. Ireland and Scotland; France extends to Wales and most of England; Scandinavia takes in the east coast; Belgium, Holland and Germany just touch London. In practice the spillovers could be considerably greater.

The area of the footprint is determined by the design of the satellite, its power, which is derived from the sun, and its position in space relative to the earth.

The cost of a communications satellite is in the order of £100 million, and the launching a further £40 million. Electronic developments may reduce costs by possibly fifty per cent by 1985, and launching costs will also drop fifteen to twenty per cent when the American space-shuttle service is in regular operation. The shuttle will permit the launching of equatorial satellites in the northern and southern latitudes, simplifying the political problem of finding countries on the equator willing to act as hosts for land-fired rockets.

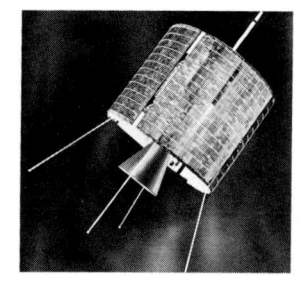

The primary use of satellites in American television is for the national distribution of over forty programme services. These range from a few hours to twenty-four hours a day. Local television and cable stations select what they please by directing their dish aerial at the appropriate satellite. Some distributors offer their services free; others recoup the costs by including advertising or by charging the local station a small fee per viewer. It is a 'wholesale' operation by programme distributors, enabling the local station to 'retail' the selected material with local advertisements which help fund their part of the operation.

The next step will be for the programme distributor to go for direct 'retailing' into the home, cutting out the retailer rather like mail-order book clubs. Each customer will need a dish aerial the size of an umbrella on his roof or in his garden, accurately pointed and tuned to the appropriate satellite. Japan was the first country to develop experimental direct broadcasting by satellite (DBS) in 1978. In the United States the Satellite Television Corporation has plans for a three-channel DBS service, to start in 1984-5 with a combined output of 400 hours a week. The consumer would pay the equivalent of about £10 a month plus the initial cost of his dish aerial and tuning equipment, said to be in the order of £200. When DBS is introduced in Britain in 1986 it may be possible to rent the additional equipment for as little as £1 per month.

In 1981 the United States, Canada, USSR and Indonesia were operating eight satellites for domestic television. Europe had been lagging behind. The French and the German governments are committed to a joint satellite project for launch in 1983-4. The first stage of the project was the successful launch of the Ariane rocket, putting up two satellites, in 1981. L-SAT, another European project, will be launched in 1985. A third multinational project is NORDSAT, shared between the Scandinavian countries, for possible launch by 1985. There are also entrepreneurial projects involving equipment manufacturers, broadcasters, publishers and financiers, for example, Radio Télé-Luxembourg. All DBS projects will have to persuade potential viewers to invest hundreds of pounds on new equipment to receive their services. They would also require a converted receiver if the television systems differ, e.g. SECAM and PAL.

Meanwhile there are two European experiments under way. 'Satellite Television', a British company, is testing an initial two hours a day satellite service, for relay by cable and television stations, supported by advertising for international products. In May 1982 thirteen member organisations of the European Broadcasting Union (EBU) started a series of one-week experiments for a pan-European service to start in the mid-eighties.

It is important to note that within the satellite business television is only one of many interests sharing a 'common carrier', and will have to compete for satellite time. For communication satellites are also used to carry other forms of traffic of varying 'load'. One hundred and twenty-five telephone conversations, for example, equal the 'loading' space of one colour television signal. International computers have been linked by satellite for many years and other business uses continue to expand. In 1980 the *International Herald Tribune* began transmitting its Asian edition from Paris by satellite for printing in Hong Kong. Transmission time is about six to seven minutes a page and it costs the newspaper around £10,000 a month to rent the satellite. The opportunities are limitless. Even prayers are now sent every Friday by satellite from mosques in Mecca and Medina for use by Islamic countries.

CABLE TELEVISION

Television can also be received by cable. Cable distribution in Britain dates back to the end of the nineteenth century when people in London could listen to West End plays in their own homes. Nowadays over two and a half million homes receive the three television networks by cable. The installation charge is £10 and a rental of £1 a week. In the last ten years there have been experiments in community programmes and in the release of feature films on a pay-tv basis. Both attempts failed on financial grounds and lack of government support. The present government is anxious to bring about a vast expansion of cable television in Britain by 1986, to coincide with the start of DBS television. The new network would have at least thirty channels and would cost £2.5 billion to cover fifty per cent of homes.

Belgium is top of the world cable league with over sixty-four per cent of TV homes subscribing to cable, the largest company, 'Coditel', claiming over 100,000 subscribers. The connection charge is the equivalent of £25, with £1 a week rental. At the present time the system offers a choice of sixteen channels in some parts of the country: four Belgian programmes, Luxembourg, three German networks, three French channels, two Dutch and three British networks.

In the United States cable television started in the late forties, with community antenna television (CATV) providing television services to small communities and urban apartments in poor reception areas. In time they moved into good areas by offering distant stations which could not normally be received. Simple alternative programming came next, paid for by advertising, followed by better programming, such as feature films and special sporting events, funded on a pay-tv basis. Individual systems were combined, linked by satellites; multiple-system owners began to buy into new lucrative areas, and with increased investment the industry further expanded.

Cable television in the United States now has over twenty million subscribers, a quarter of all television homes. Each pays a connection charge of the equivalent of about £10 and a monthly subscription of £5

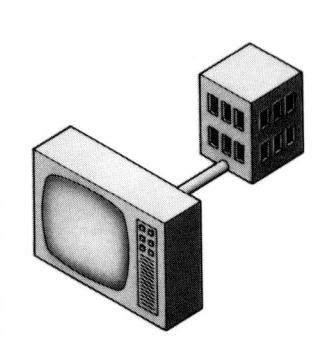

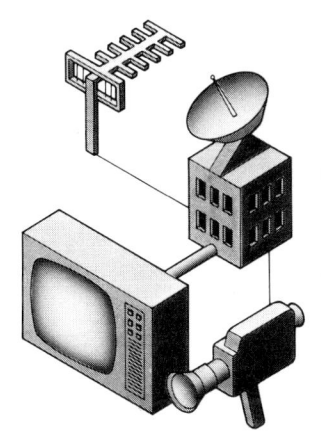

A cable station relays local television transmissions, satellite services and locally produced programmes.

for twenty channels or more. This is based on a combination of the available six or so affiliated or independent local television stations, plus community programmes and local language services, the balance being made up of programmes selected from the forty services distributed by satellites.

Of existing American cable installations, half have the capacity to carry up to seventy-two different channels on the one cable, using 'frequency' separation which works with a normal receiver. In the United Kingdom many of the installations can carry only four, or at the most eight channels, since they employ 'space' separation and require a modified receiver. The next generation of cable will be based on fibre optics, using light signals along hair-breadth glass fibres, half a dozen bunched together to form a cable equal to twenty-five existing cables and capable of carrying hundreds of different programmes or services. The slender cable will be more easily laid within existing underground ducts. This is of prime importance since costs of laying cable in America can be £50,000 a mile in urban areas. Partly for this reason Manhattan is the only one of New York City's five boroughs with cable and Chicago has none at all.

Under American law, all new cable systems since 1972 must have two-way capability, providing for interactive television. Two-way installations cannot be cheaply looped from house to house. They require a 'star' network, like a telephone system, each house linked direct to the exchange and capable of being linked one house to another. The best

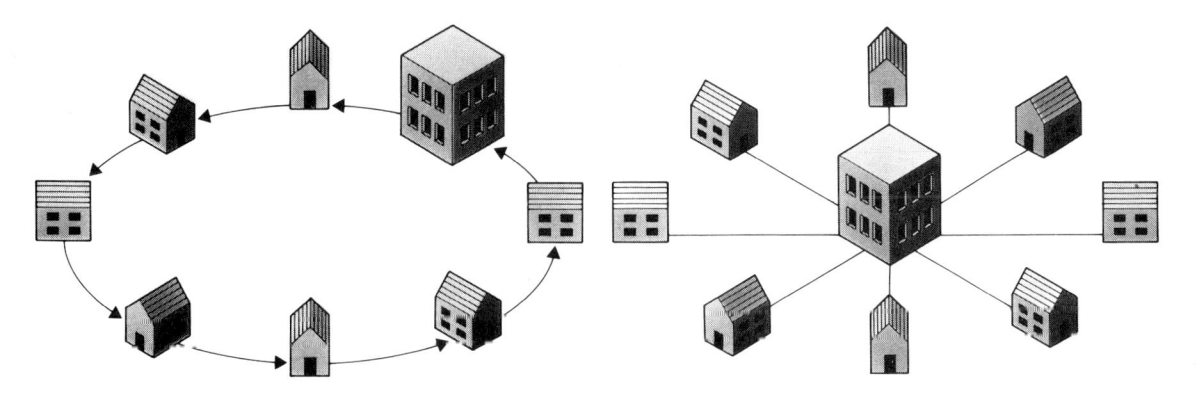

A 'looped' cable system distributes programmes from house to house. It cannot be used for two-way communication.

A 'star' cable system linking each house to a cable station for two-way communication; each house can be linked to other houses through the cable station.

known example is Warner's thirty-channel service in Columbus, Ohio, with 40,000 subscribers, known as QUBE. By means of a hand-held control unit, the viewer selects one of a variety of channels available at different costs within the three services – Premium, Community and Broadcast Television. Five response buttons are for 'polling' or reacting to given programme situations. The message light can be activated by the central computer for a variety of purposes, including security monitoring. Pilot experiments are being conducted on these lines in London and in the north of England.

Two-way installations point the way to the extension of cable for the transmission of data associated with the home, for example, emergency services (fire, police, medical, energy conservation), lighting and heating control, and the development of home terminals for banking and supermarkets, electronic mail, viewdata services, and so on. As with satellites, television is only one of many interests sharing cable as a common carrier, and, more important, sharing the costs of installation.

Forecasts in the United States are for a wired nation by 1991, with rivalry between the cable companies and the telephone companies over the design and operational responsibilities and priorities between television, domestic and commercial data transmission. To date American private enterprise has invested the equivalent of £2 billion in laying cable. To achieve half a wired nation will require some further 400,000 miles of cable which, at an average of £6000 a mile, will cost in the order of £2.4 billion. It will be seen later that the potential financial returns will justify such an investment.

Two-way cable provides for interactive television and data services in the home.

PAY-TV

Pay-tv or subscription television is the key to the financial buoyancy over the future of cable in America. Whereas broadcasting and cable services are funded indirectly by advertising or taxes over which the programme company has little control, pay-tv is funded directly by the consumer who pays for what he views. Pay-tv offers exclusive programmes – new films, sporting events and light entertainment – for four to five hours a night, so rates can be set at a profitable level. Pay-tv is now regarded by the cinema industry as a 'secondary' box office of major importance.

In America, ten million homes subscribed to pay-tv cable in 1981. Moreover, nearly a million received pay-tv programmes over the air and by 1985 an additional pay-tv audience will be established by direct broadcasting satellites. A pay-tv operator has no major capital costs for programme distribution. He buys four to five hours a day on existing cable or broadcasting stations, puts out his programme package and off-loads the costs on to the consumer who pays something in the order of £20 connection charge and about £10 a month for the exclusive programmes – which are uncensored, uncut and free of commercials.

In essence, pay-tv is an agreement between a distributor and the consumer that he will supply the exclusive programmes provided you continue to pay; if you renege, he will deny you the programmes. The deal is done at the simplest level by his disconnecting your cable if you fall behind with your monthly payments. More secure systems, essential for looped cable installations and over-the-air pay-tv, scramble the signal in a variety of ways, for example, by switching sound channels. In return for your payments, the distributor supplies you with a descrambling device, a black box that sits on top of the television set.

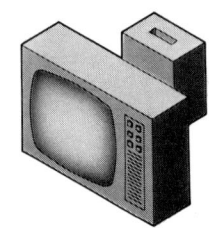

Pay-tv is based on similar programme sources as cable television, but the subscriber has to pay for the service or for particular programmes.

Far right: a British pay-tv descrambling device and selector box.

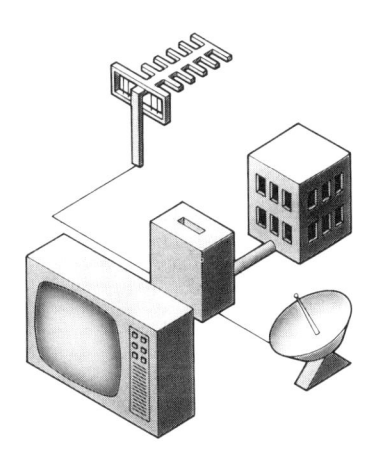

Future systems will exploit computer technology and the teletext principle of adding information to the television signal. These can be adapted for all kinds of cable and over-the-air pay-tv and for direct satellite broadcasting. In these 'addressable home terminal' systems, as they are called, your monthly payments are regularly credited by the station computer; if you fail to pay or the computer makes a mistake, as can happen, your pay-tv service is automatically cut off by remote control during frequent 'sweeps' of all subscribers by coded teletext instructions. Although requiring a substantial investment, the system enables the distributor to offer 'tiered' and 'per programme' pay-tv by switching on extra channels or extending hours for consumers who have elected to buy additional services.

Entrepreneurs in other countries are watching with interest the success of pay-tv in America. Canada already has some pay-tv but the Canadian Broadcasting Corporation is hoping that it will be run as a state monopoly involving CBC and two other main Canadian broadcasters. Australia is considering the extension of cable and pay-tv. In Western Europe, Britain is the only country involved. Seven companies have been granted two-year licences to carry out pay-tv experiments on existing cable installations. No capital is involved, though it is reckoned that it would cost £100 a household to extend existing cable installations in urban areas. Connection charges are around £30 on a simple connect/disconnect basis without scrambling devices, and about £5 a month rental. Programme choice is limited mainly to feature films.

There is little doubt that pay-tv technology could radically affect traditional methods of funding cable and broadcast television and open up new sources of finance for direct broadcasting by satellite.

HOME VIDEO

The television set can also be used to show programmes originated on video cassettes and video discs, rather like a music centre receives radio programmes and can play tapes and discs.

TELEVISION IN THE EIGHTIES

'Video' is the generic term for video cassettes and video discs used in education, in industry and in the home, together with the term 'videogram' to describe programmes distributed on cassettes and discs. Videograms are distributed physically like books and records, free of government regulation and controls, unlike broadcasting, satellite and cable television which have to share 'scarce' technical resources like frequencies. Moreover, consumers are free to play whatever programme they please, when they want.

Video-cassette machines

Video-cassette machines for the home first came on the market in 1972; but it was not until 1977 that the industry took off. A video-cassette machine works on the same principle as an audio-cassette player but records both pictures and sound. There are currently three incompatible formats: VHS and Betamax from Japan and the Philips format from Holland. Manufacturers have not yet agreed a world standard. Programmes recorded on one format cannot be played on another, nor sometimes on a later generation of machine of the same format. In 1981 VHS was ahead of its rivals in the world market.

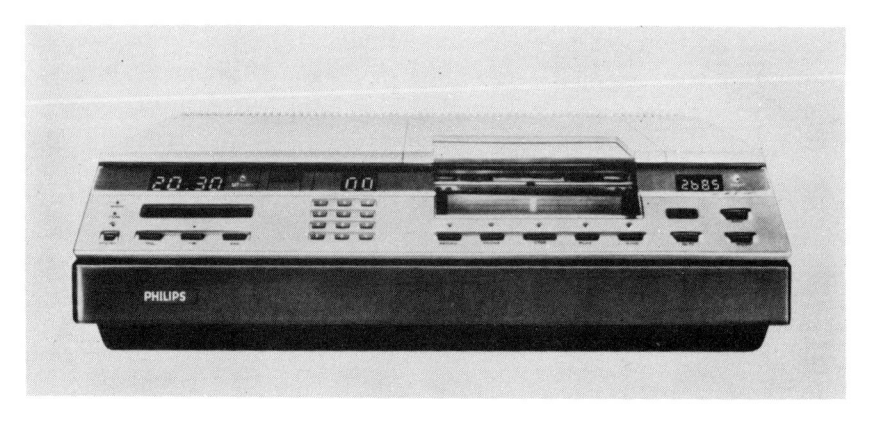

Video-cassette machines have two functions: recording programmes from the television set or from an electronic camera for 'home movies', and playing back such material or purchased prerecorded cassettes. In terms of usage, owners of cassettes in America spend on average about four hours a week recording programmes off-air, while out of the house or watching another channel, for playing back later. This is called 'time shifting'. About one hour a week is spent watching purchased prerecorded programmes, and a few minutes watching recorded 'home movies'.

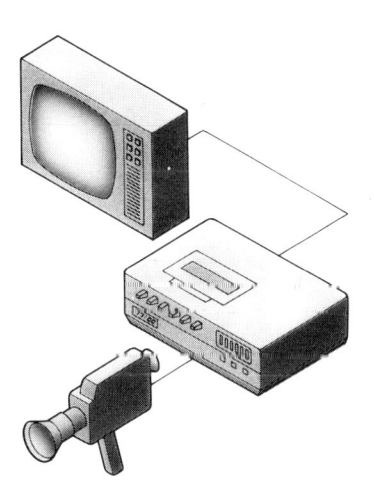

A video-cassette recorder can record and play back programmes from a television set, or from an electronic camera, and play back prerecorded material.

The simplest video-cassette machine costs the same as a colour receiver – about £500. Price depends on the facilities required. Some models can be programmed a week or more in advance to turn on and off and to change channels, or to provide a still frame and slow motion in replay.

In the United Kingdom, where over sixty per cent of television sets are rented, a video-cassette machine can be hired for about £20 a month. Blank tapes cost on average £10 for a three-hour cassette. Prerecorded cassettes cost up to £40 for a major feature film, although in Britain they can be rented for as little as £1.00 a day. Pirated cassettes of inferior quality can be bought illegally at a fraction of the price, but the industry is hot foot to bring offenders to court under international law.

Video-disc players

The video-disc player is relatively new on the market. By 1983 three different domestic video-disc systems will have been launched in the United States: Philips' Laservision from Europe, RCA's Selectavision from America and JVC's VHD from Japan. Others are still at the development stage: Sony in Japan, whose system is compatible with Philips' Laservision by agreement, and Thomson in France which is based on the same technology. The TED system from Germany is no longer a contender, although it was on the market in the 1970s.

Basically video-disc players are of two kinds: optical and mechanical. The optical systems use a laser to modulate light on discs spinning fifty times faster than an audio LP. Philips, Sony and Thomson are optical systems. The mechanical system is, in effect, the extension of the audio record player, by increasing the speed twenty-five times to include the picture as well as the sound. RCA, JVC and TED are mechanical systems.

Players of either system can only play back; they cannot record. Each system plays about an hour a side. Some systems, notably the Laservision, have stereo sound or two languages. The system can also carry teletext captions if required. Optical systems generally can run in varying degrees of slow motion, to and fro. They can also hold individual picture frames rock steady, and any of the 54,000 numbered frames on each side can be recalled at random in seconds. Mechanical players do not normally have these facilities, so they are cheaper, but suffer wear, unlike optical systems. The Philips' Laservision players cost about the same as a colour TV set; the RCA Selectavision a little less.

The JVC video-disc player from Japan.

THE TELEVISION SET

Above: the Philips' 'Laservision' video-disc system developed in Holland.

Above right: the RCA 'Selectavision' video-disc player from America.

Which system succeeds will depend on costs, design and reliability, and on the range and quality of the programmes available. The long term aim of each company is to match the price of video discs to the top selling price of quality audio LPs. When this is achieved, the pirating of programmes by copying on to cassettes will become uneconomic since the cost of the blank tape will be greater than the prerecorded video disc.

Cassettes versus discs

In the long term disc systems will be cheaper: the player is simpler mechanically, and thousands of discs can be stamped out a day. Video cassettes have to be copied in real time, which by its nature is inevitably a slower process, though expensive high-speed copiers are appearing on the market.

Optical-disc systems have the added attraction of being interactive, unlike video cassettes. The viewer is able to 'talk back' to the programme. This facility can be developed for educational purposes in terms of programmed learning and for enhancing the enjoyment of entertainment programmes. Video discs also have an important potential in industrial, legal and medical work in the storage of information, on the 54,000 numbered frames on each side. They can also be linked to computers. MCA in America have developed a player based on the Laservision principle for use in the industrial and educational field.

Nonetheless, as with cassettes, all three domestic systems are incompatible. A further complicating factor for both cassettes and discs are the incompatible television systems in the world, NTSC, SECAM and PAL. This inhibits the free flow of players and discs and cassettes and keeps prices high. This was the situation in the audio record industry in the fifties: if CBS Records with the confidence of a good catalogue had not launched the present-day LP we might still be buying 45s and 78s.

TELEVISION IN THE EIGHTIES

THE FUTURE TELEVISION SET

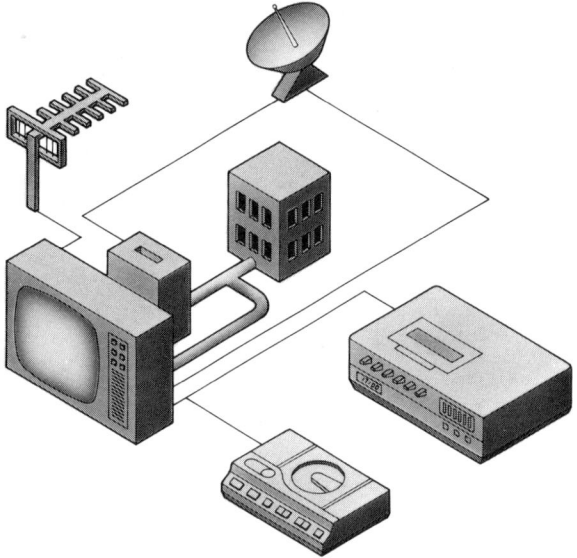

In the future a viewer will have the choice of broadcast television, DBS, two-way cable television, pay-tv, as well as using his video-cassette or video-disc machine.

The future TV set will have to be capable of receiving broadcast television and teletext by an aerial system; cable television through a 'star' network if two-way communication is required; a dish aerial in the garden to receive direct broadcasts by satellite; a decoder box for pay-tv whether by cable or over the air; a video-cassette machine for 'time shifting'; and video-disc player for access to new kinds of interactive programmes. Other services will also be available which will require the use of the television set.

Video games are the natural descendants of the electronic machines in the amusement arcades. In the mid-seventies, hard on the development of the dedicated microchip, world markets were flooded with a variety of ball games with sound effects and on-screen scoring, selling at less than £50. Subsequent developments based on microprocessors resulted in a new generation of video games. Instead of having a range of games built directly into the console, the microprocessor can be programmed to take different games simply by inserting a small cartridge.

At the top end of the market are the traditional board and card games – such as backgammon, chess, draughts and bridge – which with the aid of a microprocessor enable one person to play against a computer. The games are not cheap, though they are the most popular. Video bridge costs around £300, backgammon about £150. Chess varies between £60 and £250, the more expensive version having a voice synthesiser which calls out the moves and checkmate. In the more educational field there are the teaching games. The Race Teacher, an early example, was a programmed calculator in the shape of a racing car which set maths problems – at a cost of around £15. There are also others that are

THE TELEVISION SET

Above: a collection of video games, with *Space Invaders* on the television screen. Above right: a two way video game being played in the home.

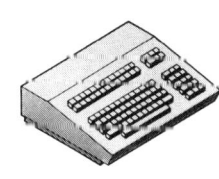

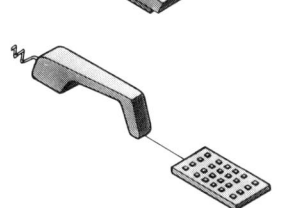

more games of logic and enable one person to play against the machine. There are also the 'reaction games' which test the operator's manual dexterity. Some are self-contained, but most benefit from display on the television screen.

A basic home computer that plugs into the television set costs £50 in Britain. It is, in fact, the core of a computer system, with add-on items that increase the memory up to sixteen times. The basic unit has its own software programmes recorded on cassettes and subjects include games, education and business/household management systems. More ambitious systems sell at around £350. Home computers extend the use of the television set as a word processor and enable the viewer in his own home to communicate with friends through their TV set, with shops, with banks, in fact with practically any source that is computer-based.

The actual means of communication is via the telephone system. The internationally agreed generic name is 'videotex', though individual countries have tended to go their own way for historic reasons or because of translation problems. In Britain two-way interactive telephone systems are called 'viewdata', and the particular system operated by British Telecom goes by the trade name 'Prestel' and was the first videotex system in the world. French videotex is called 'Antiope', the Japanese 'Captain' and the Canadian 'Telidon'. All function in the same way.

27

TELEVISION IN THE EIGHTIES

Top: a personal computer linked to a portable television set, being used at home.

Left and above: Viewdata in use commercially: a travel agent displays holiday schedules.

Centrally-based computers are accessed via the telephone network, and the appropriate read-outs are displayed on the television screen. The consumer is charged on a rate-card basis.

The United States has not yet developed comparable viewdata systems. Videotex services have so far been on cable television. And the cable industry hopes to reap even greater profits than those they have had from pay-tv, through selling videotex services and home security alarms and monitoring devices.

Teletext services that give the viewer one-way access to a variety of news and information captions is just developing in America. There are differences of view about technical standards and the Federal Communications Commission (FCC) has refrained from making a ruling, preferring to let the situation develop according to market forces. Meanwhile CBS has filed an application for a teletext service based on systems operating in the United Kingdom, France and Canada, hopefully to start commercially in 1982–3.

Britain started broadcast teletext services in 1973. The BBC invested £250,000 in setting up its nationwide, two-network, computerised, eighteen-hour-a-day service of news and information, at a running cost of only £22 an hour. In 1982 over 300,000 homes were equipped with teletext sets. On average people use teletext for something in the region of an hour and a half a week. Subtitles for the deaf have become an important added function. Graphics have improved considerably since the start of teletext in Britain. The next development will be 'enhanced teletext' which will have the ability to transmit on teletext still pictures of colour television quality. Furthermore, teletext technology can be applied to programmes on video disc, in the form of subtitles or pages of written information. The important fact is that viewdata in Britain and the teletext services, whether broadcast or on video disc, all subscribe to a common page specification: forty characters per twenty-four lines.

Above: a BBC Ceefax display.
Above right: an ITV Oracle display.

TELEVISION IN THE EIGHTIES

A video printer costing around £50 can provide hard copy from the home computer or from videotex services, say the recipe from a broadcast cookery programme printed out for use in the kitchen while actually cooking. The Japanese have gone further. Newspapers and magazines can be distributed electronically by broadcasting, cable and satellite systems for printing in the home, in colour if required. Signals picked up by the television receiver are stored in a teletext page memory before conversion to a format which drives the ink-jet printer. This has four heads for cyan, magenta, yellow and black. It takes two minutes to print a large A4 copy of a teletext page. In the long term this will have profound repercussions on the whole of the publishing world.

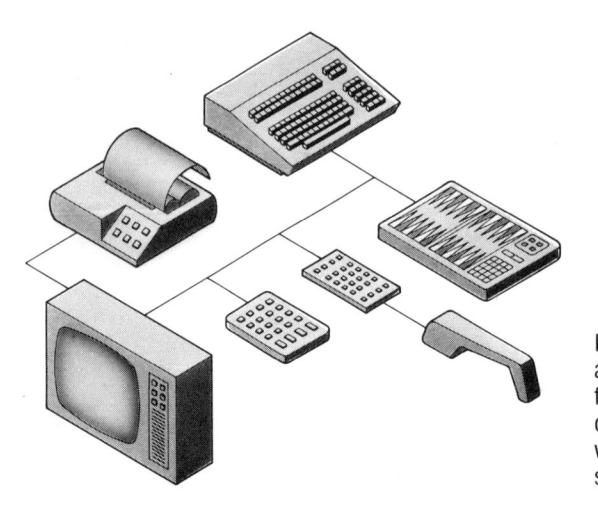

A video printer in use with a personal computer.

In the future a viewer will be able to use his television set for video games, a home computer, a video printer, as well as teletext and viewdata services.

THE TELEVISION SET

Big domestic television screens are usually three-colour projection devices. They sometimes suffer colour-registration difficulties and poor definition through enlarging a 525- or 625-line picture. The latest models provide a higher-quality resolution, a viewing angle in excess of 90° and a picture brightener which permits viewing in a normally-lit room. Large flat-screen television displays are being developed but it will probably be another decade before they can be produced in a form which would be economically attractive to consumers. By then 3D television may be possible, based on the application of holography. Meanwhile other ways of splitting the 3D images on the screen are being investigated by developing electro-optical spectacles at £100 each to replace the familiar red and green give-away plastic glasses which were developed for the 3D cinema.

Big-screen television is also used at conferences, in theatres and at sporting events.

It will be the development of high-definition television that will make the greatest impact. Sony and NHK, Japan's national broadcasting system, have agreed on an 1125-line standard which, when recorded on one-inch videotape, is said to be equal in quality or even superior to 35-mm film. CBS in America are supporting the 1125-line standard and have filed an application with the FCC (Federal Communications Commission) for an experimental three-channel, high-definition satellite system with stereo – two channels for network use, the third for feature film distribution and business use. If a common high-definition standard could be accepted by both the broadcasting and the non-broadcasting organisations, a standard would be set for world television, and the problems inherent in different line standards and colour systems would be overcome.

TELEVISION IN THE EIGHTIES

Stereo sound can be added to the television picture. Pioneered in Japan and adopted by West German television, it not only enhances the television presentation without having to make use of radio transmission, but also makes a two-language presentation possible. In Germany two-thirds of the twenty-two million households are able to pick the two-channel sound transmission. Other European countries are now actively exploring the potential of the medium for their television services.

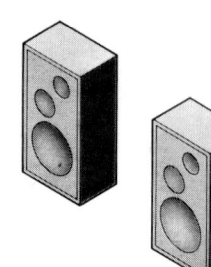

THE TELEVISION HOME TERMINAL

Enhanced television with high-definition big-screen presentation and stereo sound system can then be switched at will to any of the programme sources. It will be part of the television home terminal as it is being called.

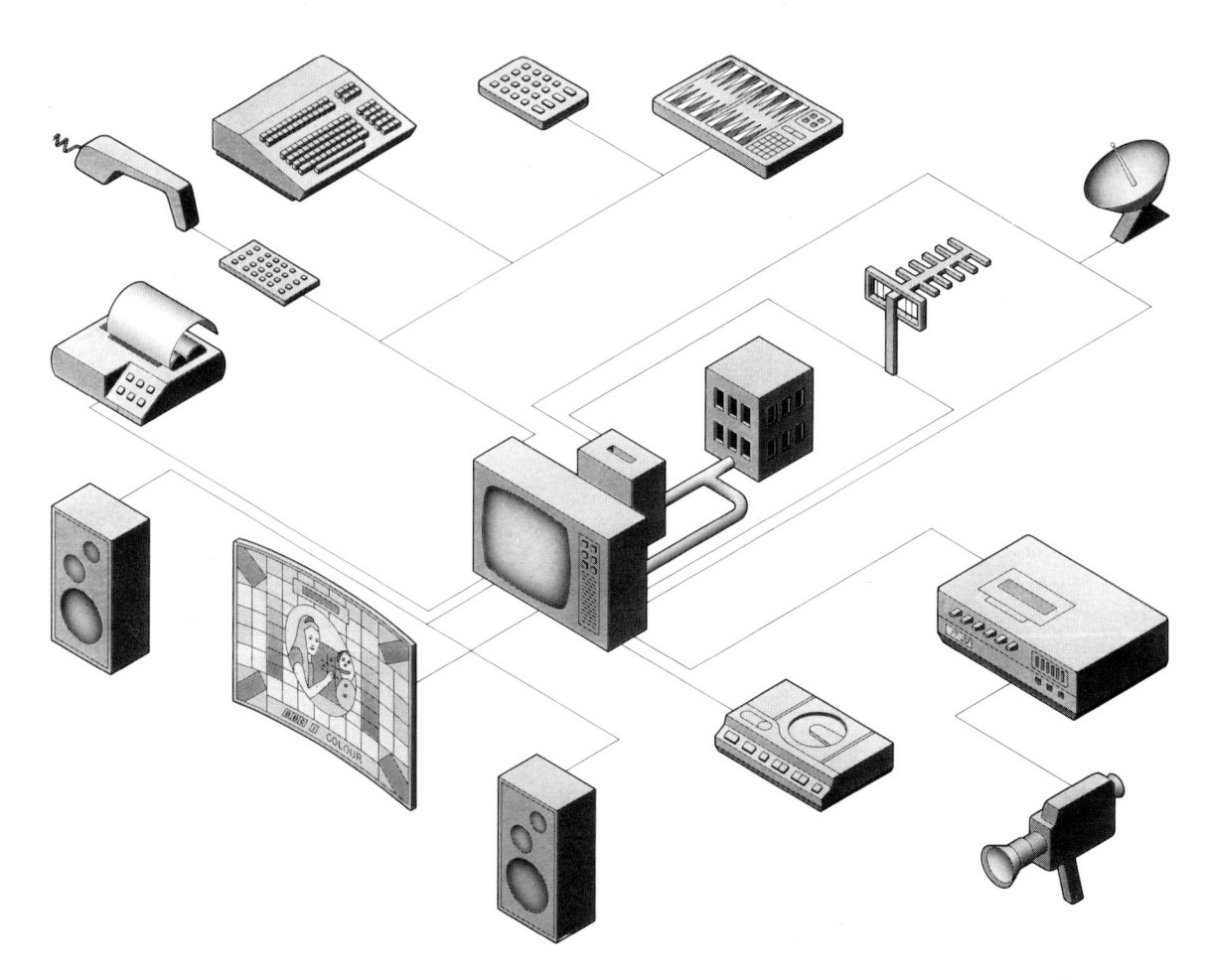

The home terminal will handle all the television programmes and services. In form it is likely to follow the modular structure of present-day audio equipment. The home terminal will, in fact, take on all the radio and audio requirements, including the idea of extensions of the various functions to other rooms, as with audio loudspeakers. It is already being planned in the United States, to be in quantity production by 1985, selling at the equivalent of £1000.

The concept of the home terminal will bring about significant changes in broadcasting, telecommunications and publishing, with implications for work and education in the home, leisure pursuits and entertainment.

The home terminal brings together programmes from broadcast television, DBS, two-way cable television, pay-tv, a video-cassette machine with its electronic camera, a video-disc player, and a variety of other services: video games, home computer, video printer, teletext and viewdata with big screen and stereo sound.

2 THE NEW TELEVISION MARKETS

The future home terminal will have two different functions: the provision of services – video games, home computers, viewdata and video printers – and the provision of programmes and entertainment from many sources – broadcasting, cable, pay-tv and home video. The service functions will have little in common with the programme functions beyond the fact that they will be in competition for the viewer's time. The programme functions, on the other hand, will compete amongst themselves not only for the viewer's time, but also for programmes and for money. It is this competition for programmes and money that forms the basis of the book. The service functions will only be considered in this context.

With traditional broadcasting the viewer has to accept the programme patterns set by the broadcasters. However, with the multiplicity of new programme sources, the viewer is becoming a consumer, with the option to pay for additional facilities to widen his choice of programmes. The only common factor is his television set.

The consumer's choice is in proportion to the amount of money he spends. In broadcasting, for a given television station with a running cost of, say, £3 a month, the consumer's choice is nil. He has to be party to whatever programme pattern the broadcaster presents. In cable, for £5 per month, the consumer may have a choice of twenty or more channels. The pay-tv consumer pays £10 per month, as the domestic satellite user probably will, for the added attraction of exclusive programming. With home video the consumer's choice is theoretically infinite, outside news and topical programmes, but he has, of course, to pay £500 for the video equipment and up to £40 per prerecorded programme, and even £10 for a blank cassette to record his own programmes.

Individual consumers make up an audience. The nature of the new audiences is determined by the amount the consumer is prepared to pay, and what degree of choice or exclusivity he wants. Broadcasting, which is the cheapest system for the consumer, with little choice, works in audiences of millions, and cannot deal with small minority interests. Cable, though more expensive for the consumer, offers a wider choice and works as a system with audiences of hundreds of thousands, and can deal with minority interests. Pay-tv, even more expensive since the consumer has to pay extra for exclusivity, works with audiences of tens of thousands. Direct broadcasting by satellite, still in its infancy, should work with audiences similar to pay-tv. DBS will have the capacity, theoretically, to

THE NEW TELEVISION MARKETS

	Broadcast Television	Cable Television	PAY-TV	DBS	Home Video
Capital Costs A consumer with a television set wishing to extend his choice beyond broadcast television has to pay installation costs for the additional services	None	£10	£20	£200	£500
Running Costs The consumer also has to pay for the programmes directly or indirectly	£3 per month	£5 per month	£10 per month	£10 per month	£40 per programme
Consumer Choice The number of programme channels available is in proportion to the amount of money the consumer spends. With broadcast television the consumer has to accept the schedule	Nil	20 channels +	20 channels ++	6 channels	Infinite
Audience Size Broadcast audiences are measured in millions; cable television in 100,000s; Pay-TV in 10,000s; DBS, initially, also in 10,000s; in Home Video the audience for one programme can be measured in 1,000s.	Millions	100,000	10,000	10,000	1,000
Market Value	£3 million	£500,000	£100,000	£100,000	£40,000

Relating running costs to audience size, a broadcast station will have £3 million for every million viewers per month. With home video one programme could produce £40,000.

expand its audience size to 100 per cent penetration. Video is the most expensive source of programmes for the consumer, but enables him to choose whichever programme he wants, when he wants. It is a highly selective system and should enable small audiences of thousands to buy and watch the programmes of their choice, no matter how specialised the subject-matter.

Relating audience size to running cost, it is possible to establish some kind of market value for each system. In broadcasting, each million viewers would pay £3 million a month. In cable, with its smaller audiences, each 100,000 viewers would pay £500,000 a month. Similarly, with pay-tv, each 10,000 viewers would pay £100,000 a month.

The differential is best seen with video where a potential viewership of 1000 would pay up to £40,000 for one programme. No broadcasting organisation in the world can afford to spend £40,000 on a minority audience of 1000. It is obvious when comparing broadcasting to 'narrow casting' – pay-tv and video – that the opportunity for providing exclusive programmes for specialist or minority interests becomes financially feasible.

It is not easy to imagine, from 1985 onwards, the consumer faced by an almost impossible set of choices. In the United Kingdom, where one-quarter of television homes have two or more sets, the average viewer at present sits in front of his set for four hours a day. In the United States, where nearly every other home has two or more sets, the average viewer watches for six and a half hours a day, and can choose between a wider range of programmes, distributed by broadcast transmitters, cable, pay-tv, subscription television, satellites, and video cassettes and video discs. In fact, a viewer living in Sunningvale, north of Baltimore, could have the choice of 200 feature films in any one week, including local cinemas.

How far will the liberated viewer use his initiative in the future and seek out programmes not readily available on existing systems, and go out and buy or rent a video cassette or disc? Will he be happier to choose from what is available at the flick of a switch between broadcast and cable television and the pay services? How many consumers in urban areas will actually set up a satellite dish aerial in their garden? How many will prefer to have their domestic and foreign television programmes fed over existing installations? Some of the answers can be drawn from the phenomenal rates of growth of the new systems, their projected incomes and their estimates of programme demands over the next five to ten years.

3 GROWTH OF THE MARKETS

Whilst Japan leads the world technologically, the United States has the lead in the exploitation of cable, pay-tv, video and satellites. It provides, therefore, a good yardstick for Britain and other European countries for the next decade. It has to be stressed that there are fundamental differences in the quality and type of television programmes in Europe and the United States, and that which meets the needs of American society may not appeal to Europeans. Furthermore, the degree of government regulation is much greater in Europe than in the United States, and this will be an important factor in determining the rate of development of the new systems in Europe.

In the United States the new systems have grown beyond all expectation in the past three years. One of the many factors contributing to their success has been the availability of alternative television at a time of growing dissatisfaction with the range and quality of the commercial networks. Consumers were prepared to pay to watch films unedited, uncensored and free of commercials. In addition, aggressive marketing of the new systems capitalised on the American's interest in new technology and his ability to pay for it.

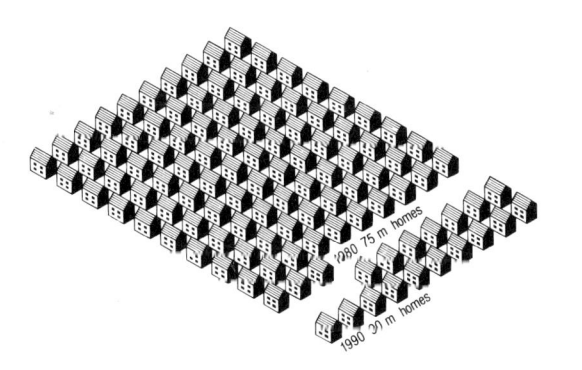

United States 1980-90.
Television homes in millions.

The television market, however, has almost reached saturation point in the States. In 1980 there were seventy-five million homes using television, and it is forecast that by 1990 there will be a maximum of ninety million, almost 100 per cent of all homes. The new systems will be competing for their shares in this television marketplace.

At the end of 1980 there were almost twenty million cable subscribers in the United States, twenty-five per cent of homes using television. This represents a fourfold increase in ten years and, if the

37

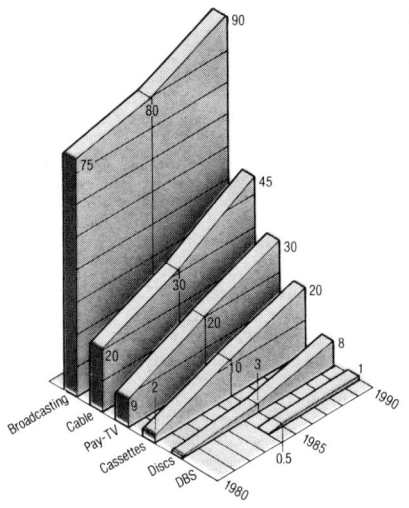

United States 1980-90. Growth of new systems into television homes, in millions.

present trend continues, it is estimated that there will be at least thirty million cable subscribers in 1985, rising to forty-five million in 1990, fifty per cent of all television homes.

The recent growth of pay-tv has been even more phenomenal, particularly since 1978. In 1980 this had reached almost nine million pay-tv subscribers, twelve per cent of homes using television. By 1985 it is expected that pay-tv will have penetrated twenty-five per cent of television homes, twenty million subscribers. At the end of the decade the figure might be as high as thirty million, one-third of television homes.

Video-cassette machines first appeared on the domestic American market in 1977. By the end of 1980 the machine population was approaching two million, a three per cent penetration in four years. Estimates vary widely about future growth. The general consensus puts the cassette-machine population at ten million by 1985, rising to twenty million in 1990. The first video-disc system, Laservision, was introduced in America in late 1979. With the entry into the market of two new systems, Selectavision and VHD, estimates are in the order of a population of three million players in 1985, rising possibly to eight million by 1990. Direct broadcasting by satellite did not exist in the States in 1980. By 1985 the system is planned to be working there in one form or another. Optimistic estimates for direct reception in the home put a DBS population of half a million in 1985 rising to one million by 1990.

These growth rates have financial significance. The incomes of the new systems are rising at rates proportionate to their growth. Broadcasting, on the other hand, has already reached near-saturation and its income curve is flattening out. Despite the increase in the number of homes using television to ninety million in 1990, the income growth for the commercial networks will not really be affected, since revenue comes mainly from advertising and rates will not rise in real terms because actual audience sizes are not going to change radically. Moreover, the new systems will be making inroads into the traditional sources of funding for broadcasting.

In order to bring out comparisons of the potential incomes of the new systems the following formulae have been used. The 1980 rates for cable have been taken at the dollar equivalent of £60 per annum, pay-tv at £120 per annum. It has been assumed that owners of video-cassette machines will spend £100 buying four prerecorded programmes a year, whilst owners of disc players will spend £120 buying ten discs a year. The same figures could apply to renting cassettes or discs. DBS payments have been assumed to match pay-tv rates. All forecasts have been made at 1980 prices.

Right: United States 1980-90. Income forecasts in £ millions.
Far right: United States 1980-90. Theoretical projection of demand for programmes.

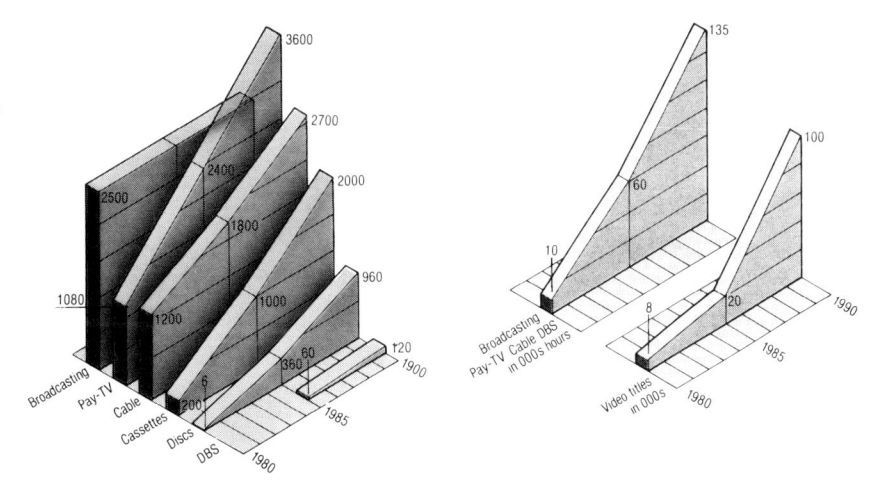

As the sum of money spent by consumers on the new systems increases, there will come a time when the overall income will match, and may finally overtake, the total incomes of the broadcasters in America, undermining the traditional patterns of the funding, the production and the distribution of television programmes. What will these forecasts mean in terms of programme demand?

An early American forecast was for a sixfold increase in programme demand by 1985, grouping broadcasting, cable and pay-tv together. In 1980 10,000 hours of original programmes were produced, excluding news broadcasts, so the forecast for 1985 could be 60,000 hours, and for 1990 135,000 hours. This should also meet the requirements of DBS in 1990. Video programmes will probably increase from the 8000 titles known to be on sale legally in America in 1980 to somewhere between 15–25,000 in 1985 on a guesstimate basis, and to 100,000 by 1990.

America, with the world lead in the exploitation of the new systems, may well be the pattern for development in Europe. For example, by 1980 the Western European market for video-cassette machines had already exceeded the American domestic market, despite the differences between European television standards. Moreover, more prerecorded video cassettes are now being sold in Western Europe, despite the national and cultural make-up of the different countries.

Looking specifically at the United Kingdom, broadcasting is in eighteen million homes, third in the world in percentage of penetration after Japan and America, possibly rising to twenty million by 1990.

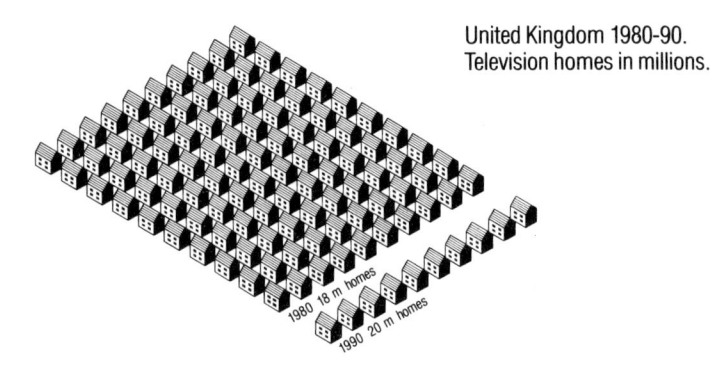

United Kingdom 1980-90.
Television homes in millions.

Cable is in just over two and a half million homes, with the possibility of doubling by 1990 or before with little capital cost, since existing cable installations could easily be linked to an additional two million homes. If Government plans go ahead, though, half the population could be linked to a national cable network within the decade. Pay-tv is currently the subject of a two-year Government experiment, starting in 1982 and involving approximately 100,000 potential subscribers. If this is successful, pay-tv could then be extended not only to existing subscribers, but also to the proposed national cable network.

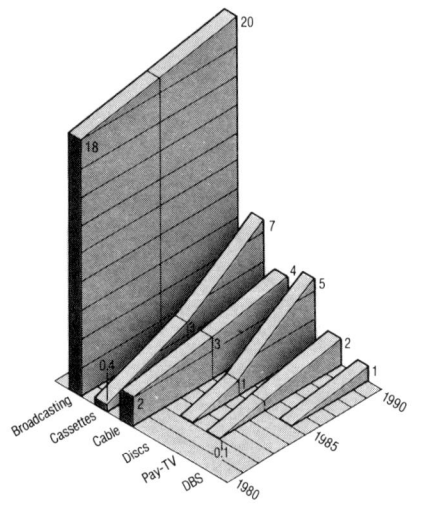

United Kingdom 1980-90.
Growth of new systems into television homes in millions.

In 1980 there were already 400,000 video-cassette machines in the United Kingdom. By 1985 this could have risen to three million or more.

GROWTH OF THE MARKETS

The forecast for 1990 is in the order of seven million machines. In 1980 video-disc players had not yet arrived in the United Kingdom. By 1985, when all three systems should be on sale, there could be over one million disc players in the country. By 1990, there could be as many as five million, of different formats.

When the British proposals for two satellite services come into operation in 1986, apart from amateur enthusiasts there might only be a few homes equipped with dish aerials and tuners; but it is thought that there could be as many as one million by 1990.

These growth rates for the United Kingdom do not yet pose the same financial threat to broadcasting organisations as in America. Although broadcasting in the United Kingdom reaches nearly 100 per cent of homes, the hours of transmission are by no means approaching saturation within the twenty-four hours as in America. This factor will enable commercial television to increase its income from advertising by the introduction of breakfast television in 1983. Channel Four will be overlapping with existing hours of broadcasting and will be competing with ITV for available advertising money. The BBC, on the other hand, can only increase its income by the Government raising the annual licence fee. This means that overall the income for broadcasting in the United Kingdom will level off, as available transmission hours are filled.

In order to bring out comparisons of the potential income, the same formulae have been used as for the American situation though the costs are different. The 1980 rates for cable have been taken at £12 per annum and pay-tv at £60 per annum. Owners of video-cassette machines will spend £120 buying four prerecorded programmes a year, and owners of disc players will spend £150 buying ten discs per annum, or the equivalent in renting programmes. DBS payments will compare with pay-tv rates. All forecasts have been taken at 1980 prices.

Right: United Kingdom 1980-90. Income forecasts in £ millions.
Far right: United Kingdom 1980-90. Theoretical projection of demand for programmes.

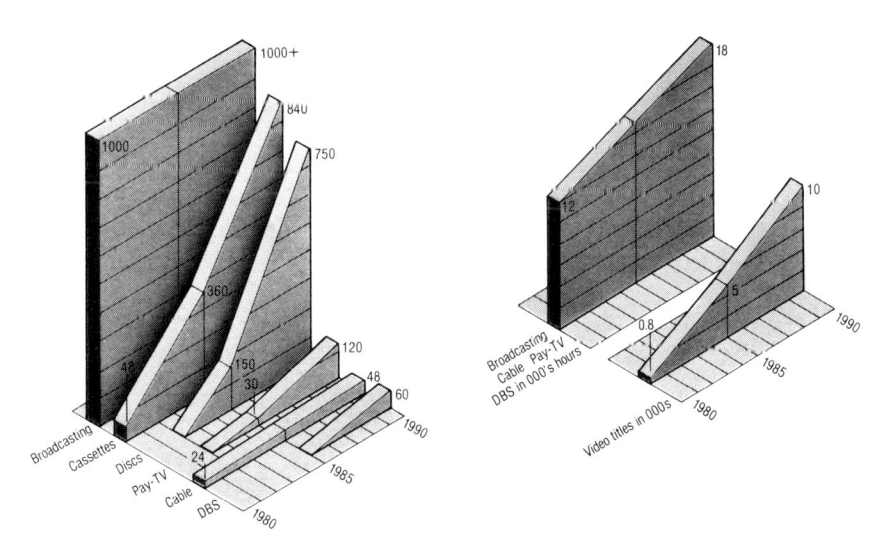

What will these forecasts mean in terms of programme demand in Britain? Certainly the broadcasters' programme needs are going to increase. In 1980 the BBC produced 6500 hours of original programmes, including news broadcasts, and ITV 5500 hours. By 1985 Channel 4 will have increased the demand by a further 4000 hours a year. Two breakfast television services, BBC and commercial broadcasting, could account for an additional 2000 hours a year. The requirements of pay-tv and satellite television in Britain will make little impact initially, though there would be a growing demand if cable were extended. In 1980 800 video titles were on sale officially, published in catalogues not pirated. This could rise on a guesstimate basis to 5000 by 1985, probably doubling by 1990.

Looking now at the other European markets, there are already television sets in almost every home, although the funding of broadcasting varies from country to country. There are already extensive cable networks in Belgium and Holland although none of these have yet been developed for pay-tv systems. Both the French and German Governments have been forward-looking in the development of satellites for television transmission; services should be in operation for 1985. Scandinavia have also planned to start a Nordsat satellite system for Norway, Sweden, Denmark and Finland for the mid-eighties. There are other government plans involving Britain, Switzerland and Italy, as well as private projects involving manufacturers, broadcasters, publishers and financiers. In the video market the ownership of cassette machines throughout Europe is growing rapidly, especially in Germany, Holland, France and Scandinavia. Together they form a larger home video market than the United States.

Adding the demands for programmes in the United States to the future demands in Europe leads to the key question of this book – who is going to produce programme material on this scale? The established sources in a given country – the broadcasters and the film industries – or new sources – newspaper and magazine publishers, book publishers, record producers, the theatre, and educational and other institutions?

First, some attempt at an analysis of the types of programmes required by each of the new markets needs to be made, before trying to answer questions about possible new sources of programmes.

4 ANALYSIS OF THE NEW MARKETS

The established providers of programmes for broadcast television throughout the world are the national broadcasters themselves and the film industry, as well as some independent producers. In Europe the broadcasters produce the bulk of their own programmes. In America the commercial networks are mainly responsible for the production of news, information and sport; most entertainment programmes are bought in, chiefly from Hollywood. The question is whether the traditional programme makers will also be able to meet the different requirements of the new markets.

By the nature of their mass audiences, broadcasters have to schedule a wide variety of programmes – news, features, films, sport, children's programmes, light entertainment, education, drama, religious programmes, and specialist services for schools and universities. Commercial networks aim to place programmes with the most appeal during prime viewing time to gain the biggest audiences to satisfy their advertisers. Public service broadcasting is expected to carry some serious programmes of less general appeal, but still catering for millions. In America the three commercial networks and their affiliates dominate broadcast television. CBS, for example, devotes about thirty per cent of its programming to news, information and sport, twenty per cent to prime-time entertainment, about ten per cent to late-night entertainment, and the balance to daytime programming. PBS, the publicly funded network, makes a different contribution and provides a service for more specialist interests.

The broadcast viewer has no choice of programmes on a given channel; he has to accept the broadcaster's schedule. The cable viewer has his choice extended – he can choose from relays of the networks, local stations and locally produced community programmes and access programmes. The local programmes will range from earnest features of one kind or another to the most risqué access programmes, for example, in New York regular interviews conducted in the nude, men and women sitting comfortably on a hearth rug in front of a blazing fire. For the remainder of channels, the cable operator, wherever he is, will select a dozen or so of the forty programme services distributed by satellite to the whole nation. Most are 24-hours-a-day services: four news channels, one sport, one weather, three religious, one black, one Spanish, and one Jewish; three popular music channels, three cultural channels, specialist channels for women, children and health; and 'super-stations' such as

TELEVISION IN THE EIGHTIES

```
CHANNELS LISTED IN                    All programs are in color ex-
NEW YORK METROPOLITAN EDITION          cept those designated by (BW)
BROADCAST STATIONS

NEW YORK CITY           MONTCLAIR, N.J.        BRIDGEPORT
(2)-(53) WCBS (CBS)     (50) WNJM (Ind.)       (49) WEDW (PBS)
(4)-(67) WNBC (NBC)
(5)-(64) WNEW (Ind.)    NEW BRUNSWICK          SATELLITE PROGRAM SERVICES
(7)     WABC (ABC)      (58) WNJB (Ind.)
(9)     WOR (Ind.)      (See listings on Ch. 50.)   (ESN) ESPN (Entertainment and
(11)-(73) WPIX (Ind.)                                     Sports Programming Network)
(13)-(75) WNET (PBS)    NEWARK                 (MSG) Madison Square Garden
(25)    WNYE (PBS)      (47)-(62) WNJU (Ind.)  (USA) USA Network
(31)    WNYC (PBS)      (68)-(60) WWHT (Ind.)
                                               PAY-TV SERVICES
GARDEN CITY             PATERSON
(21) WLIW (PBS)         (41) WXTV (SIN)        (HBO) Home Box Office
                                               (SHO) Showtime
SMITHTOWN               HARTFORD, CONN.        (WHT) Wometco Home Theatre
(67) WSNL (Ind.)        (3) WFSB (CBS)              (Subscription TV)

                        NEW HAVEN
                        (8) WTNH (ABC)

Symbols for hearing-impaired viewers:
(CC) Closed-captioned    (OC) Open-captioned        (S) Interpreted in sign language
(Special decoder needed) (Visible without decoder)      (Interpreter appears on screen)
```

The wide range of television channels available in New York, published in *TV Guide*.

A selection of American cable and broadcast programme guides; some are given away, some are sold.

WTBS, Ted Turner's 24-hour independent channel from Atlanta, Georgia, which provides family entertainment based on feature films and sport. These cable services come free to the viewer, supported for the most part by advertising.

If the viewer subscribes to pay-tv, his choice is extended further. Over-the-air subscription television is limited to a nightly package of feature films. Pay-cable, fed from satellites and other sources, offers the viewer the latest feature films, special programmes for children, serious music programmes and pop, relays of opera, ballet and plays and so-called adult entertainment consisting of varying 'degrees' of pornography. Programmes are unedited, uncensored and free of commercials.

The significant fact is that, with the rapid rise in the number of channels available, cable is increasingly moving into generic scheduling following the pattern of radio broadcasting. Television broadcasting, by comparison, has very few outlets and still has to base its appeal on mixed scheduling.

The way cable and pay-tv have developed side by side is seen best in the two-way cable station, QUBE, in Columbus, Ohio, which was started in 1977 by Warner Communications with thirty channels, divided into three groups. There were nine 'Premium' channels charged on a pay-per-view or a pay-per-day basis; ten community channels; and the existing television channels and other information services – all for a basic subscription fee of £6 per month. The nine premium channels included first-run feature films, 'Movie Greats' from the past, classical and contemporary 'Performance', 'Better Living', 'How To' courses, sports, special events, various interactive 'Qube Games', College at Home and adult films. Five response buttons enable the viewer to react to programme situations, in sport, for example, talent contests and quiz games. They are also used for test-marketing of new products and for reaction to political speeches. The community channels included a 'Columbus Alive' channel, 'Pinwheel'-devoted to non-commercial, non-violent children's programmes, religion, audience participation programmes, and other information services, four of which were teletext channels. The ten 'television' channels included the three commercial networks and their affiliates, the local educational station, three independent television stations, a public access channel and a teletext programme guide. For an extra fee, the Qube subscriber can have five 24-hours-a-day stereo music channels. This basic pattern has been continually updated to meet changing audience requirements and the need to remain profitable.

In systems of this kind the cable operator can log the viewing pattern of each household and this led to some anxiety about security, particularly when it was revealed that a well-known local personality had a predilection for the adult movie channel.

As early as 1980, the video-cassette owner in America had the choice of some 8000 different prerecorded cassette titles. By far the largest number were feature films, ranging from classic favourites such as *The Sound of Music* and *African Queen* to later additions such as *Saturday Night Fever* and *Jaws*. The balance was made up of pornographic films, from the cinema and other sources in America and Europe. Now feature films are released for sale or rental on video cassette almost at the same time as a film has its cinema première. There are indications that other types of programmes are becoming popular, serving educational and specialist interests such as leisure pursuits. Video-disc programmes, unlike video cassettes, have so far to be manufactured by the machine-makers, and have concentrated on the feature film and specialist-interest market, which particularly exploit the interactive features of the Laservision machine.

In Europe the pattern of broadcast programmes differs from country to country, although interest in news, sport and feature films is common to all. Existing cable networks mainly relay domestic broadcast programmes, or transmissions from adjoining countries. Unlike America, language is a formidable barrier.

In the United Kingdom, the programme choice on the pay-tv experiments is almost exclusively restricted to feature films; the licences granted by the Government have been issued on the basis that neither advertising nor sponsorship can be included and no exclusive coverage of any event may be shown. In other words, the programmes offered cannot be obtained in direct competition with the existing broadcast organisations. There is little doubt that the development of cable television, based on thirty channels and available in half the homes in Britain, would have profound effects on the acquisition and scheduling of programmes. It could lead to 'generic' television, national channels distributed by cable for local relay, not by satellite as in America. These could include a news channel, a sports channel, a weather channel, a 'pop' channel, a music and arts channel, an educational channel, and others. New organisations could come into being to provide these generic services.

As far as video cassettes are concerned, the majority of programmes on sale are feature films, as in America, although there have been a number of attempts at original programming to widen the range; for example, gardening, motor racing, beauty culture, opera and ballet. In 1981 feature films still accounted for over half of all home-video releases in the United Kingdom, with so-called adult films accounting for twenty per cent. Children and family programmes accounted for five per cent, pop music was also five per cent and sport eight per cent. Serious music and the performing arts accounted for nearly three per cent.

Video-disc catalogues will also be based on feature films, with some experimental programmes exploiting the unique features of the Laservision system. One such example is a BBC programme about garden

Right: some British video catalogues. Feature films dominate the market.

Intervision
NEW RELEASES
A RIP ROARING BLOOD AND THUNDER EPIC
The Exterminator
ZULU DAWN
MAGNETIC VIDEO
A TWENTIETH CENTURY-FOX COMPANY
Tora! Tora! Tora!
VIDEO FROM Thames Television International
VCL Video Catalogue
FROM THE PROFESSIONALS at PINEWOOD
IVER FILM SERVICES LTD. present
VIDEORAMA
Take away Movies on Video.
United Artists
AUTUMN 81/SPRING 82
THORN EMI VIDEO
PRESENTS
FLASH GORDON
THE ELEPHANT MAN
THE JAZZ SINGER
LIFE OF BRIAN
THE DEER HUNTER
EXPLOSIVE VIDEO ENTERTAINMENT from DERANN
THE DAY THE EARTH CAUGHT FIRE
SCREAM FOR VENGEANCE
ONE MAN JURY
DEATH OF A SNOWMAN
THE LIFE TAKER
THE BOLDEST JOB IN THE WEST
LUST FOR REVENGE
WHITE COMANCHE
PANIC CITY
GODFATHER OF HONG KONG

birds, which includes some 170 pages of additional teletext information about the seventy-five birds featured in the programmes as well as teletext subtitles for the deaf.

Some types of programmes are common both to broadcasting and to the new markets, specifically feature films. The film industry has already increased its revenues by releasing its classic films. Now productions are budgeted, taking into account the potential earnings from distribution into the new markets. For example, the film *Rocky* recovered all its production costs by exposure on pay-tv in one year. As far as video cassettes are concerned, Magnetic Video, an American company, made the first deal with a major film company, Twentieth Century-Fox, in 1977 to distribute fifty feature films on video cassette. Following their success, they are now a part of Twentieth Century-Fox itself.

Other major film distributors have followed suit. However, they do not release major new feature films into the mass-broadcast market first; they go for the smaller markets initially – pay-tv and video – where the consumer will pay for exclusivity; in fact to video first, to pre-empt the pirating by recording off a pay-tv showing. Pay-tv follows video, then selected cable outlets. Broadcasting follows at least three years after the video and cinema release. This is called sequential distribution. Walt Disney are the masters at this strategic planning. They bring out their classic films, *Snow White, Bambi,* and so on, every seven years for worldwide distribution to catch each new generation of children. They are also adept in exploiting the cable and broadcast television markets with their adaptations and compilations, though they did not enter the home video market until 1981, and then only on a restricted rental basis, so as to keep control of their product.

Other types of programming demanded by the new markets are in the areas of sport, children's interest, light entertainment and education, traditionally elements of the broadcasting schedules. The question is whether the broadcasters will be able, in terms of resources, money and within their constitutions, to follow the pattern of the film industry in releasing programmes from their libraries to the new markets as well as distributing new programmes. This may well produce problems for the broadcasters, if they follow the economic argument of sequential distribution, in terms of releasing programmes into the new markets before they have been broadcast. For example, a major opera or ballet, with international artists, or some other 'definitive performance', might be broadcast first to persuade video owners to buy a recording. Sport, on the other hand, might go on pay-tv first and be released in different versions for the other markets. In the near future events like the Open Golf Championships may be available all day on pay-tv with the highlights of the day's play on broadcast television or cable in the evening. At the end of the contest, video cassettes would be on sale.

Because the new systems have an increasing demand for programmes, and since they can also cater for very small audiences profitably, new types of programmes are emerging. These are in areas that the traditional programme providers – the film industry and the broadcasters – cannot, by the nature of their economics, cover. Inevitably this is providing not only opportunities for independent producers and distributors, but also opportunities for other industries, totally unrelated to broadcasting, to adapt their crafts to the growing needs of television.

These other industries include newspaper publishers, book publishers, and data services, the record industry, and possibly radio. They are taking advantage of these new opportunities, but they are also adapting the new technologies to suit their own traditional activities. Moreover, artistic, entertainment and other institutions are considering ways of exploiting their own activities in the television market to their own ends. These include sporting organisations, theatres, art galleries, museums, educational and religious organisations, as well as industry and commerce.

The significance of these developments is that the traditional programme makers – in particular the broadcasters – are being challenged by new programme makers. New sources of programming and services are coming from other industries and related institutions who want to exploit television in its broadcast sense, to diversify and to widen their spheres of influence – and to make money. This will inevitably raise questions of funding, resources, structures, legal constraints and government regulation.

5 NEW SOURCES OF PROGRAMMES AND SERVICES

RADIO

Television was the natural offspring of radio. It inherited techniques of programme production and scheduling, and continues to share programme ideas, writers, artists and musicians. Where radio and television have remained within one organisation, economies have been possible by sharing management and editorial services. It is not always so easy to marry the two operations. They have each developed different technical skills and resources and tend to operate on different time-scales. A news item can be simply phoned in by a journalist anywhere in the world and be broadcast directly by radio. For a television news item, camera and sound equipment, a small transmitter or even a satellite feed would be required at considerably greater cost in time and money.

In the United Kingdom, on the BBC national networks, radio broadcasts three times as many hours as television. Radio has over thirty hours of news in a week, against eight hours on television. Radio 4 alone carries seventy hours a week of current affairs, features and documentaries, against thirty hours on television. On the other hand, television has twice the sports coverage of Radio 2, the sports network, and nearly four times the amount of programmes for children than Radio 1, the young people's network. In light entertainment the two carry the same amount. Radio 4 originates twice as many hours a week of drama as television, at a fraction of the cost. And taking music of all kinds, radio broadcasts around 350 hours a week, nearly 100 times more than television.

Because radio broadcasts for many more hours than television, it has developed 'generic radio', where each station has its own distinctive style and format. In America there are eighteen such generic categories. The list speaks for itself: easy listening; adult contemporary; classical; country and western; black; disco; middle-of-the-road; modern country; all news; rock; Spanish; news/talk; variety; standard/nostalgia; religious; golden oldies; soft contemporary; and album-oriented rock.

Altogether there are nearly 9000 radio stations in the States. New York has 120 stations. On average there is one radio station for every 25,000 citizens, compared with one for every 250,000 in the United Kingdom. There are technical limitations on any similar development in Europe. Reception on medium wave is difficult at night and there is the prospect of increasing interference ahead because of the limitation of the waveband. Moreover, stereo sound cannot be accommodated on the medium wave. The only way out is to exploit the VHF waveband.

Whereas America already uses the full band allocated by international agreement, Government regulations in Britain restrict the broadcaster's use of the VHF band. There is agreement to release the band gradually, but not totally until 1996. Government 'emergency' services are occupying the VHF broadcast-frequency allocation, despite alternative allocations in other parts of the spectrum. Once the handover is completed, all radio stations in Britain – National, Regional and Local, whether commercial or public service – will be on the one VHF waveband. Station identification will be simpler and most radios will have push-button tuning. This is important because research shows that on average a radio listener stays tuned for eighteen hours before changing station. With television, which has push-button selection, channel switching is commonplace.

The quality of sound broadcasting is continually being improved – first stereo, then quadrophonic sound. The emphasis has now shifted to 'digital' sound which can be carried over great distances with no loss in quality. For instance, in June 1981 the first concert from Shanghai to be transmitted digitally by satellite was received in Britain with spectacular results.

Radio is also developing more information services. In America, for example, some of the National Parks have their own VHF stations which carry a continually repeated recorded commentary describing the park as you drive through. The particular frequency is advertised on road signs as you enter the park. In Britain, the BBC is developing a medium wave system called Carfax which will provide drivers with traffic information through a small electronic adapter attached to their car radios,

Carfax: the improved radio service for motorists.

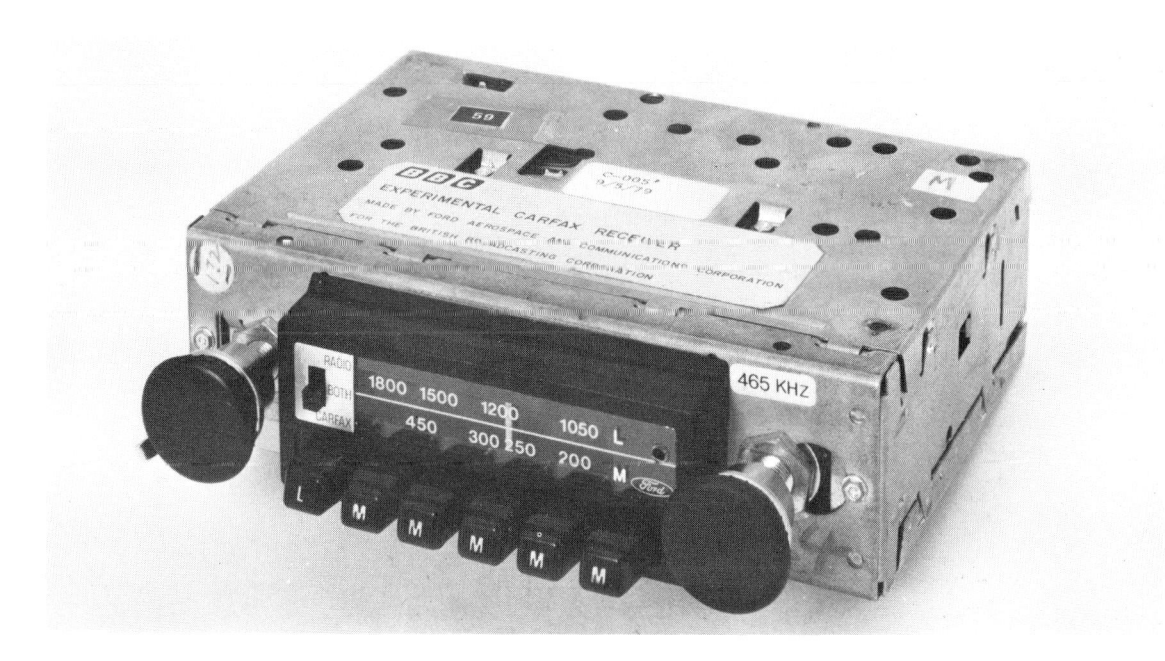

while they are travelling anywhere in the country. The information being fed to the driver will only be what he requires on his own particular route. At present a driver has to listen at all times for intermittent traffic news. When Carfax is introduced, the driver need not necessarily be listening to the radio at all; the signal will be automatically received, the radio will be activated, and the spoken information passed to him. The Dutch and the Germans are pursuing their own systems, and hopefully a European standard will be adopted.

The BBC and broadcasters in Sweden and France are also experimenting with a new system called 'Radio-data'. Although primarily intended as an aid in identifying stations when tuning, Radio-data will also be able to carry other information such as programme or music titles, or sports results. This could be a visual display on a small electronic screen like a digital clock or a voice synthesiser which could give the same information. EBU discussions are taking place to agree a common European standard.

Radiovision has for many years been used in the United Kingdom to describe certain radio educational programmes illustrated by film strips. In 1981 the BBC conveniently took over the term to describe a programme broadcast simultaneously on television and radio. The requirement is that the broadcast sound, without pictures, should be comprehensible. BBC Radio frequently broadcasts the stereo sound of a television opera or concert simultaneously. In America there have been experiments in broadcasting a foreign-language soundtrack of a film on radio simultaneously with the English version on television, rather than providing subtitles.

Radio-data: a radio receiver equipped with a simple visual display.

Radiovision could be more than that. It could be, for example, a neat and economic way of providing a breakfast television service by building on an existing radio programme. It would be a new animal, neither conventional television nor conventional radio. Its unique advantage is that the audience – whose listening and living habits are very different in the early morning – could wake up and listen to the radio, go downstairs to watch television and drive off to work, still listening to the same compatible programme. Even with the appeal of many breakfast television shows in the United States, one in four Americans listens to the radio in the morning, and of those with breakfast television switched on, 56 per cent only listen to the sound.

The concept of Radiovision illustrates the strengths and weaknesses of both media, but neither has the quality of permanence of a newspaper or a book. News announcements or sports results can only be heard at the time of broadcast. The development of teletext is changing this. Information, supplied by the television station, can be called up by the viewer when he wants it. At the local level it could well be the radio station which supplies the service. Local television in the American sense is too expensive in Europe, whereas local radio stations can be run comparatively cheaply. A radio station could fill several pages of local teletext news and information, possibly paid for by local advertising.

Radio could also publish its programme schedules on teletext. Musical scores could be shown during radio concerts, or the words for community hymn singing. Cookery recipes could be shown and be printed out on a home printer. Telephone numbers and addresses given in programmes could remain on teletext for some hours for reference. Where applicable, teletext services of this kind could attract advertising.

In the longer term, radio might be broadcasting not just the words and music as now, but computer programs and other abstract forms of information which could be recorded on $\frac{1}{4}''$ tape or cassettes. These could then be played into a television set in association with the home computer and video home printer, part of the television home terminal.

Given this kind of radio development in association with television, and the proliferation of generic programming and specialist services available on VHF radio and its cost efficiency, radio organisations see a more encouraging future than in the sixties when television seemed set on dominating not only radio, but the cinema, the theatre and newspapers, magazines and books. In fact, radio and television have become competitive, particularly in daytime hours. While television attempts to appeal to the broadest possible audience, radio has, for some time, reacted to the challenge by making its appeal more specific. This enables programme makers and advertisers to identify audiences precisely by age group, interests and spending bracket. Moreover, the cost of advertising on radio in America has risen only sixty-five per cent since 1967, half the television rise of 127 per cent. Cost effectiveness puts radio in a strong position to consider exploiting its interests in the new markets, and, in commercial radio, to do so with profit.

RECORDS

The record industry holds an exclusive pool of musical talent, in both the pop world and classical music. It exploits this talent through the sale of LP and single records on a worldwide scale and on internationally accepted technical standards. It also allows both radio and television stations to broadcast its records. The output of some radio stations is entirely records. In Britain, however, record companies place a restriction, called 'needletime', on the number of hours that records can be broadcast. The broadcasters have to pay an agreed amount for each 'needletime' hour. There are no such restrictions in America. The extensive use of records by radio is a key factor in promoting record sales and has become an integral part of marketing strategy.

A similar relationship has been developed with the film industry. Nowadays, especially with musical films, record distributors arrange for the release of LPs to coincide with the premières of films. *Saturday Night Fever*, for example, sold thirty-six million copies worldwide. Television shows are similarly exploited. The title music of the successful series *Brideshead Revisited* sold as a single for 50 pence. There was also an LP of Sir John Gielgud, who was one of the actors, reading extracts from the original book. The title music for a series about Lloyd George has earned something approaching £50,000.

Nonetheless, the record industry has its problems. Overall sales have dropped in the last three years. This is due in part to the general world recession, but also to piracy and counterfeiting, and the increase of home taping. It is now possible to buy a pirated audio cassette of most popular LPs for a tenth of the market price, particularly in countries in the Middle and Far East. Moreover, counterfeit records are sold at full price, but without the record company receiving any of the proceeds.

The practice of home taping is growing, especially since more highly sophisticated equipment is widely available. Friends will share LPs with each other, taping copies on to blank tape at a quarter of the cost of buying the LP itself. Once again the record industry loses out. People also tape records directly from the radio, particularly in the pop world, though broadcasters try to discourage this by talking over the start and end of records. There is so far no technical solution to prevent either this or the copying of LPs. In the United Kingdom record manufacturers have been pressing for a levy to be raised on the sale of equipment and blank cassettes, but this has so far been rejected by the Government in its 'Green Paper' on the reform of copyright published in July 1981. There is a levy on the sale of recording equipment in Germany, for example, and on the sale of blank tape in Austria.

The industry is continually improving the technical quality of its products. Stereo sound is now the accepted world standard. There comes a time, however, when any additional improvement will only give a marginal return, since it will involve the consumer and the broadcaster in buying additional equipment.

The record industry now has the opportunity to exploit its musical talent in the making of programmes for the new television markets. Traditionally television broadcasters produced their own music programmes; it was not economically viable for the record industry to compete in this area. Twenty years ago, it was only possible to see pop groups like the Beatles at live performances, planned in association with the release of their records or on an occasional live television appearance. Television could not, however, re-create the unique 'sound' of the new groups since they did not then have the facilities of the recording companies. The practice of 'miming' to records developed and proved to be a cheap form of presentation, albeit slightly dishonest. In fact, some broadcasters improved their facilities and insisted that pop groups should mime to a prerecorded television version of the sound.

By the mid-seventies the record industry recognised that pop music on television was going to demand more than pictures of groups singing and playing. Television producers experimented with more dramatic lighting effects and unusual camera angles. Singers no longer performed in a studio or theatre, but were seen singing outdoors in natural surroundings. In the pop world itself a new industry sprang up outside television broadcasting, in Los Angeles, in London and in other pop capitals in Europe. Small independent television producers were hired by record companies to film groups at concerts and in other locations. The film was edited together with imaginative visual effects for broadcast use or was transferred to video cassette. These became known as 'promo' cassettes since their principal purpose was to promote the sale of audio records. They were played in record shops and department stores, and projected on big screens at discos. The pop 'promos' were also regularly screened on television shows that exploit the record 'charts', like the weekly series *Top of the Pops* in Britain.

In the early days the 'promo' cassettes were sometimes distributed free as part of the marketing operation. The situation is changing as more people buy video-cassette players, and video versions of an audio record now have a market value. Indeed, it was only a matter of time before a series of 'promo'-type presentations of one group were linked and sold as a video album in its own right, released on the same day as the audio LP. The first of these was *Eat to the Beat* by Blondie in 1980. The video album costs £29.95 and the LP £5.00. The impact, too, is quite different. On the Blondie video album, the order of tracks had to be changed to take account of the visual impact.

At the end of 1981 there were around 100 pop video programmes on sale in Britain. 'Live' concerts dominate most catalogues. Some of the most popular releases are basically collections of 'promo' tracks. The latest Olivia Newton-John, *Physical*, cost £250,000 to produce and is also a video disc with stereo sound. It was the first complete video album to be broadcast in its own right by the BBC, with no form of presentation between the various tracks.

With the television experience of the pop world behind them, the record industry has moved into video versions of classical music recordings, in competition with established broadcasting organisations. Each Christmas for many years the BBC broadcast a television relay of the traditional Carol Service from King's College, Cambridge. In 1979 EMI produced their own television presentation with stereo sound. They expended more time and effort than a broadcaster can usually afford in producing a timeless and definitive interpretation of the Carol Service. In terms of distribution, EMI took a leaf out of Hollywood's book, and sold the videogram production, as they call it, to the BBC, who broadcast it twice, and to broadcasters in nine other countries. It has been distributed on video cassette and is going to be released on video disc with stereo sound. In October 1981 EMI followed this with the simultaneous release on video cassette and on a digitally recorded disc and audio cassette of Beethoven's Violin Concerto, priced at £34.50 for the video cassette and about £5 for the audio versions.

As the new television markets develop, other original videogram productions are following. Initially these will be more concerned with 'definitive' performances in concert halls, in opera houses and with ballet companies, in the tradition of filmed presentations. But as confidence grows, the industry may turn to the creation as opposed to the performance of music, on the lines developed by the BBC with its series of 'Master Classes' with great artists like Paul Tortelier and Elisabeth Schwarzkopf.

It is unlikely that the record industry will experiment outside their fairly well-defined musical areas. With their experience in producing pop 'promos' they may lead the way in experimenting with the visual presentation of music of other kinds: for example, seeing the hands and faces of the performers in chamber music; serious music heard to the accompaniment of visual images, both real and fantasy; and new kinds of presentation of rock concerts for big screens with 3D holographic pictures and stereo or quadrophonic sound.

The record industry starts with the advantage in the serious music world of having contracts with artists and musicians which can be extended to include video rights, just as has been done in the pop world. They also have two other advantages. Firstly, their extensive distribution outlets in High Street shops, department stores and mail-order clubs throughout the world; already video cassettes are stacked alongside audio cassettes and discs. The same outlets can develop rental systems as

Right: some original productions from the British record industry for the home video market. Both audio and video versions have identical packaging design.

Christmas Carols from Cambridge
ITZHAK PERLMAN
BEETHOVEN
VIOLIN CONCERTO
CARLO MARIA GIULINI
PHILHARMONIA ORCHESTRA
EMI
HIS MASTER'S VOICE
DIGITAL
ITZHAK PERLMAN
CARLO MARIA GIULINI
BEETHOVEN: VIOLIN CONCERTO
PHILHARMONIA ORCHESTRA
Olivia Physical
Olivia
Physical
Olivia Newton-John
Physical
Featuring her new album "Physical"
+HOPELESSLY DEVOTED TO YOU
+A LITTLE MORE LOVE +MAGIC

well as outright sales. The second advantage is in the very name, video disc. Consumers will automatically associate video discs with record companies.

The production of 'video' music and its distribution puts the record industry in competition with broadcasting, the traditional provider of music on television. In the pop world, broadcasting has already lost out. Radio relies heavily on LPs and singles in its pop output; television pop programmes already rely on the 'promo' cassettes made and supplied by the record industry. Similar competition could arise in serious music. The record companies could become a new source of programme material for television, cheaper for the broadcasters to buy than they can produce themselves. In Europe some broadcasters may take the initiative and go into partnership with a recording company – the broadcaster providing the television skills and resources and the recording company the artists and the performance, each sharing in the profits from worldwide sales in the new markets. Irrespective of the means of production, as more 'video' music becomes available music of all kinds will form a bigger share of television output.

This has already happened in America with the increasing number of outlets in cable television. In August 1981 MTV, the first round-the-clock music channel, began. Each hour, linked by a DJ, consists of about three-quarters of an hour of music, five minutes of news and interviews and eight minutes of advertising. MTV is distributed by satellite to over two and a half million cable homes. Accompanying stereo sound is broadcast on FM radio channels. Another satellite music channel, planned for July 1982, is WBLS, 'the World's Best Looking Sound', with a projected two million subscribers. It will provide urban, contemporary music, rhythm and blues, pop, disco, jazz and gospel from a major New York theatre, twelve hours every day. It will be supported by advertising.

Later in the year, the Heartbeat Media Network plans to start a daily six-hour music programme, 'Music for your Eyes', also supported by advertising. It will be based on rock, soul, jazz and other formats and will make use of video discs for the first time. At the beginning of 1983, the Nashville Network will start a twelve-hour-a-day country entertainment service, made up of a six-hour package of music, and a country music-based soap opera screened at 6 pm and midnight. This will be distributed by satellite and will be supported by advertising and sponsorship.

Of the established cable systems, the ARTS Network distributes a three-hours-nightly programme based on the visual and performing arts to over six million homes. CBS Cable, another cultural channel, has three million subscribers for its daily twelve-hour schedule of concerts, ballet, jazz and drama. Both these are supported by advertising. RCTV's Entertainment Channel, which starts in 1982 on a pay basis, will also include music of all kinds. Bravo, another pay-tv network with 120,000 subscribers, also includes concerts and operas in its ten-hours-a-day schedule.

Not just broadcast television, but cable and pay-tv and later DBS are going to rely on the record industry as an important programme source. In the pop world the record companies will be expected to provide promotional video and film clips free of charge, just as review copies of audio discs are distributed. MTV, for example, is reported to have acquired 400 such items. Profit will come from the subsequent sale of video cassettes or video discs to the public. The cultural cable channels will rely heavily on European broadcasters for music and opera and ballet programmes. RCTV, for instance, has negotiated American rights to BBC productions, which will comprise about forty per cent of its schedule. The cultural channels will also turn to record companies for serious programmes. EMI have foreseen that demand in their videograms of King's College Carol Service and the Beethoven Violin Concerto. Programmes of this kind based on a 'definitive' performance will have a long and profitable life on broadcast television, cable, pay-tv, DBS and home video, not just in one country but throughout the world.

CINEMA

In its heyday the cinema was an unrivalled source of entertainment and a great money spinner. Most towns in Britain in the 1930s could boast of at least three cinemas. Films were exclusive to one or other cinema chain, identified by the studio credit: the roaring lion for MGM, the Morse transmission for RKO, the striking gong for Rank.

Each studio had its exclusive stars under contract, although fashions changed. During the second half of the thirties child stars topped the popularity polls, first Shirley Temple, then Mickey Rooney. Whereas in the thirties every other star was a woman, the proportion dropped to one in ten in the seventies. Stars have remained a key factor in film production, distribution, publicity and cinema exhibition. Throughout the sixties, though, it was the director's 'artistic vision' that mattered. By the seventies this had shifted to consideration of a film's 'economic profile'. As far back as 1915 the issue of 'art form versus industry' was significant enough to be brought before the American Supreme Court, which ruled unanimously that films were only an item of commerce. It was not until 1952 that the Court revised its decision and declared that 'motion pictures are a significant medium for the communication of ideas'. Thereafter, the cinema was accepted as an art form and an industry.

Although remaining an effective art form, the cinema is no longer a highly successful industry. The exception is India where it is still booming – 596 cinemas were opened in 1980, bringing the total to around 11,000 serving a population of 680 million. Elsewhere the industry is static or in decline. In Britain attendances, which were reaching one and a half billion a year in the fifties, fell to 100 million in 1980, ten per cent down on the previous year, and they fell a further ten per cent in 1981. In America they numbered about seventy million per month during 1981, nearly ten million a month fewer than the previous year.

The reasons are easy to find. People prefer to watch films, albeit a few years old, in the comfort of their own home on television, with usually a much better supporting programme than the cinema offers. It is also cheaper. The price of cinema tickets rose nearly twenty per cent in Britain in 1981; the cost of one ticket for a West End cinema is now about four times the weekly cost of the television licence. Moreover, since suburban cinemas are generally the first to close down, people have to travel further to see a film.

These factors have affected not only the number of people who go to the cinema, but also the make-up of the audience. The 15–17 age group are now the most frequent cinema-goers in the United Kingdom, followed by the 18–24-year-olds. The over 45s are bottom of the list, preceded by the 35–44s. This has affected the choice of films and the kind of advertisements shown. Some advertisements even cost over £100,000 to produce, but they are guaranteed a defined, captive audience.

Independent producers argue that, by limiting the choice of films to escapist entertainment and blockbusters appealing to this narrow section of the market, distributors and cinema exhibitors have excluded large sections of the potential audience and a source of revenue. With a more representative choice of film a night out at the cinema could regain some of its former magic. The most recent example is the award-winning film, *Chariots of Fire*, about the 1932 Olympic Games. This was produced in the best tradition of British film-making; it is an intelligent film with a strong narrative, appealing to a wide audience. Against a budget of just over £3 million its estimated income is around £8 million (estimated at £2m from British cinemas, £1m from British TV and video, £3m from foreign cinemas, and £1.5m from foreign TV and video).

One of the first companies to recognise that, while cinema audiences were falling, the overall market was certain to expand was Pearson Longman, the British publishing group who own *The Financial Times*, the *Economist* and Penguin Books. They formed a film-financing partnership, Goldcrest Films International, with the National Coal Board Pension Fund, Electra House and a number of investment trusts. Goldcrest funded the initial script development of *Chariots of Fire*, which was then bought and produced by Twentieth Century-Fox in partnership with Allied Stars, a new finance company formed by United Stars Shipping.

The funding of films in Britain and the future of the industry generally has been under review by an Action Committee appointed by the Government in 1976. Under the chairmanship of Sir Harold Wilson, the Committee has published a number of reports, recommending the setting up of a British Film Authority, tax concessions for investors and changes in the law to extend the levy raised on cinema exhibition to showings on television. The Committee also say that the three-year ban on television screenings should be imposed by law, and that the television industry should help support a renamed National Film and Television School. They have also drawn attention to the potential use of satellites and cable for the distribution of films. Britain is, in fact, the only European country with no state-regulated system for channelling television money into film production. ACTT, the trade union covering both the film and television industries, wants the government to put up £5–£6 million a year and establish a British Film Authority, to replace a number of separate bodies. Top producers would like to see a figure of £50–£80 million raised annually from a levy on television showings, maintaining that this could at last open the doors and give access to City financing on a proper footing.

Meanwhile in America, fewer films are being produced. By mid 1981 only fifty-three new films had been started by the major Hollywood studios, compared with fifty-eight at the same time in 1980, and many are cutting their budgets. Between 1978 and 1980 production costs rose by sixty-five per cent, increasing the average cost of feature films to £5 million, more than double the 1975 figure. Total investment in 1980 was nearly £750 million, plus £350 million for marketing. Nearly a third of Hollywood expenditure goes on selling rather than making the product.

However, Hollywood films earned £1000 million in 1980 sales worldwide, over £200 million coming from Canada, Germany, France, the UK and Japan alone. This has turned a number of American flops into successes. As far as is known, only thirty of the sixty-four feature films costing more than £5 million made since 1956 have recouped their investment within the United States. It is foreign sales and rentals, and latterly television sales, that bring many films into profit – so much so that distributors now follow a prearranged sequential marketing pattern.

It is forecast that by 1985 home video and pay-tv will rival the cinema as the number one source of income for the studios. The industry believes that cinemas will continue to exist to make a film's reputation, but mainly as promotion for home video and pay-tv screenings. In America, it is estimated, pay-tv paid the equivalent of £125 million in licence fees for feature films in 1981. Home Box Office (HBO), because of its size, has been able to negotiate a flat-rate buying policy. Others pay a rate according to the number of subscribers. At one end of the scale blockbuster films command 50–60 cents per home; at the other, 'B' pictures from independent distributors may only pull in 3–5 cents per home. Demand is so great to fill the 24-hour satellite programme schedules that major distributors are re-releasing prewar films. An even wider variety of titles will probably lead the three largest pay-tv networks to some sort of specialisation. So will exclusive buying. HBO is already buying up an estimated fifteen to twenty per cent of all new English-language cinema films released in the United States, frequently before they have been shot and edited.

Broadcast television usually has to wait three years after a new film's release in the cinema. Even with these constraints the three channels in Britain broadcast over thirty-five hours of feature films and series a week in 1981. On average they cost something less than £10,000 an hour. This is the cheapest form of television and some of the most popular, even though Cinemascope proportions are squeezed into the television screen.

Meanwhile the form of cinemas is changing. Many old-fashioned, big-screen cinemas seating 1000 or more have been converted to smaller multi-screen theatres. These bring down running costs and at the same time give more choice for the customer, which permits an increase in admission charges for specialist films, probably viewed in greater luxury. The Rank Organisation in Britain had 596 single-screen cinemas in 1950. In 1982 they had 231 screens in only ninety-four cinemas. Some spaces in the old cinemas, such as restaurants and foyers, were not easily converted into small cinemas serviced from a central projection box. This was solved by the installation of video projection systems. The first video theatre with seventy-four seats, part of a multi-screen cinema, opened in Norwich in 1978. Others have developed around Britain, either as part of an old cinema or, as in Acton, London, where the video theatre is combined with a 'pub' and the audience can order snacks and drinks. In a shopping complex in London, a video theatre has a continuous programme of cartoons to entertain children while their parents shop, reminiscent of the cartoon cinemas at railway stations. It is alongside a soft drink and snack bar, an indication of how video theatres are becoming part of general consumer facilities.

Video theatres in Britain now use front projection, with the video projector set in the ceiling. At present video-cassette machines are used; in future they may be replaced by video discs with stereo sound. The changeover from one cassette machine to another is automated, as is control of the house lights and the playing of music during intervals. The projectionist only has to load the machines. He monitors the programme remotely on a black and white television set displaying the output of a video camera in the actual theatre.

Compared with cinema projection, the picture quality in the video theatre is less bright and definition poorer: viewing from the side gives an unacceptable picture. Against these shortcomings, sound quality is better than optical sound on film. Video cassettes are ten times cheaper than 16 mm film prints, and the capital cost for a video theatre varies between a quarter and half of a comparable film installation.

The next development will be to distribute films electronically to video theatres by satellite. The video theatre could be equipped with a bigger dish aerial, capable of receiving more stations than the domestic version. This would enable the operator to present a mix of programmes rather like a broadcasting schedule, or relay one satellite service continuously. One of the British satellite channels planned for 1986 will distribute new feature films and sport on a subscription basis. In parts of Europe where there are foreign-speaking minorities, video theatres could relay the appropriate foreign satellite service. Problems of copyright and government regulation would have to be faced.

As in the development of the cinema, a number of video theatres could be grouped under one roof, offering a variety of programmes. Electronic distribution by satellite will be cheaper than the physical distribution by video cassette and by video disc. This will, in turn, affect the traditional breakdown of costs in the cinema to the advantage of all concerned. At present the exhibitor takes two-thirds of the gross box office, the distributor one-third (out of which he has to pay for the distribution rights, the making of prints, shipping and advertising).

However, the development that will really revolutionise the film industry is 'electronic production'. Television cameras will replace film cameras and video tape will replace film. The 'film' of the future will be electronically edited like a television programme. Many more facilities will be available at less cost and it will be possible to amend the script on a word processor linked to electronic editing equipment.

In order to match the picture quality of film, the Japanese have developed a high-definition television system with stereo sound based on 1125 lines instead of the American and European standards of 525 and 625 lines. They also propose to change the aspect ratio of the screen from 4:3 to 5:3. This not only conforms more to cinema-screen proportions but is an improvement in itself in terms of picture composition. Engineers and producers agree that the picture quality is comparable to 35 mm film, even when projected onto a video screen. The problem is that High Definition Television (HDTV) requires at least twice as much 'space' on a satellite or cable channel as conventional television. The same applies to recorded HDTV. Nonetheless Francis Ford Coppola, the Hollywood producer, who has already made two experimental 'films' on HDTV, has announced that all his productions will be recorded on video tape by 1983/4.

When HDTV is developed, the relationship between television and the traditional film industry could become closer if television, in the form of broadcasting, cable, satellites and home video, adopts HDTV as a common production standard, as a first step towards achieving a world television standard.

THEATRE

The cinema grew out of the theatre. It employed the same writers, actors, singers, dancers and musicians. Together they lost out to television in the sixties and seventies, but the theatre lost out most. While films made originally for the cinema became a main feature of television, theatres had to subsist on their stage performances.

In the United Kingdom, the falling turnover forced the Society of West End Theatres in London to adopt a more positive attitude towards 'marketing' the theatre. For instance, they set up a ticket booth in Leicester Square on the lines of the New York system in Times Square whereby unsold tickets of the day are offered at half price. In 1981, its first year of operation, 350,000 tickets were sold, admittedly only a fraction of the eight to nine million annual total, but many of the purchasers were in fact first-time theatregoers. Research also shows that about a third of potential theatregoers are more deterred by travel difficulties than by the rising cost of theatre seats. The Society is thus pressing for more late-night transport and cut-price packages of travel and theatre tickets.

Their efforts have succeeded in halting the declining turnover: in 1981 attendances averaged sixty per cent of capacity, just over the break-even point. Some plays and musicals do exceptionally well with regular full houses. Musicals, though they cost up to £500,000 to mount, can take well over £100,000 a week in box office receipts. However, some commercial theatres are seeking the support of the subsidised theatres in outer London and the provinces. In 1980, in fact, nearly half of West End commercial productions originated from subsidised theatres. Indeed a producer can cut his financial risk by up to forty per cent by launching his production in an out-of-town subsidised theatre.

Subsidised theatres, on the other hand, both regional and national, are also going through a difficult time because of cuts in public funding, and a number are threatened with closedown or curtailment of their activities. The situation is seen most clearly in London, where the four national theatre, opera and ballet companies face severe cuts in their Arts Council state subsidy for 1982–3. The Royal Shakespeare Company wanted an increase of thirty per cent and received just over half that. The Royal Opera House is to receive about £10 million, an increase of just over eight per cent. The English National Opera subsidy will be almost £5 million, an increase of nearly nine per cent. The National Theatre will receive nearly £6 million, an increase of just over ten per cent. The companies argue that the increases will not even keep pace with inflation, let alone cope with rising costs. Actors cannot act more to increase productivity, nor can admission charges go higher than the market will stand. With the number of seats limited by the physical size of a theatre, the outlook is bleak.

Commercial sponsorship continues to help. The National Youth Theatre, which lost its Arts Council support in 1980, has received in the order of £60,000 from Texaco, to see it through to the end of 1982. In all, sponsorship in the subsidised theatre amounts to only about one and a half per cent of the Arts Council budget.

In 1981 a significant turn of events took place. Until then theatre plays had occasionally been televised, opera and ballet more frequently. But they were costly exercises for the broadcasting authorities and to some extent were of minority interest in terms of network audiences. On 2 January 1981 the BBC broadcast a relay of *Tales of Hoffman* from the Royal Opera House which was to become the first venture of Covent Garden Video Productions. The company had been launched that day to exploit the potential of opera and ballet in the new television markets, especially on cable, video cassette and video disc with stereo, as well as in established television broadcasting and, later, in satellite broadcasting. The operation is based on a three-way contract between the Royal Opera House, the BBC and Covent Garden Video Productions, to relay a minimum of fifteen productions, both opera and ballet, over the next five years. The Opera House supplies the performance, the BBC contributes the television production skills and resources, for which it gets its relay free, and Covent Garden Video Productions pays all fees and obtains total world rights. The deal is not exclusive, though the BBC has first refusal on any production. Covent Garden Video Productions can make deals with any of the ITV companies or any other partner. Likewise the BBC can make separate deals with the Opera House. The unions involved have each reached settlements based on a percentage of gross takings.

The project was obviously a gamble since no one knew how quickly the new markets would develop or how long the company could afford to wait for a return on its investment. The outlook is promising. *Tales of Hoffman* was presold to Metromedia in the United States; it was relayed live to Austria; and has since been seen in France, Germany and other European countries. It is planned for showing on pay cable in the States on RCTV, the new up-market cable service. The Opera House stands to gain on its past earnings from TV relays and at the same time the cost to the broadcaster is reduced. This points to a new kind of partnership between the theatre and television. It also provides a new kind of investment for finance companies and sponsors.

In New York, as in London, the commercial theatre faced a decline in the seventies. Broadway theatres fought back. Investment rose from $17 million in 1976 to $143 million in 1980 and is now in much better shape. Some regional and cultural theatres are subsidised by local funds, but there is no federal funding on the scale of the Arts Council. It is television that is showing the greater interest. Stage productions are attractive propositions for the new up-market cultural cable channels and for home video. Theatres understandably welcome their interest, not only for the fees payable but for exposure of their productions to a much wider audience and to potential ticket buyers. The contracts range wider than in Britain though none is so ambitious as Covent Garden Video Productions in terms of world markets.

Showtime, a pay-tv cable company with two million subscribers, was the pioneer in 1979. They relay one play a month, including regular Broadway shows. In 1981 Home Box Office, the largest pay-tv company with six million subscribers, have also started relaying stage productions, and by 1983 will be presenting monthly stage plays. CBS Cable, with nearly three million subscribers, have similar plans. Bravo, a smaller cultural cable service with 120,000 subscribers, is also planning to add stage plays to its programming which consists largely of ballet, opera, dance and jazz. Bravo was the first company to relay performances of symphony concerts and ballet.

Another newcomer is the ARTS channel, with nearly five million subscribers. They are exploring the idea of sponsorship of cultural programmes on television and have already done a co-production of *Macbeth* with the Lincoln Center in New York. RCTV's Entertainment Channel, which starts in 1982, intends to present one major theatrical production a month, including Broadway shows. In addition it plans to produce programmes from cultural centres such as the Lincoln Center and to make approaches to foreign theatres. The PBS network has plans for a pay-tv cable system which will include stage productions. The network will encourage a partnership between the nation's cultural institutions and public (non-commercial) television stations.

There are a number of problems in turning theatre presentations into successful television. In choosing plays, for example, there has to be a strong plot which is easy to follow in order to combat the distractions of home viewing compared to the concentration of a theatre audience. There is also the question of whether a play should be recorded for television during a live performance or be transferred to a television studio with or without an audience. With modern unobtrusive television production equipment and post-production editing, there are strong arguments for preserving the theatre atmosphere, even if the recording has to be done over three or four performances to permit 'retakes' of complicated scenes. Televising from a theatre can, however, produce conflicts between camera perspectives and voice projection and may require elements of restaging. For this reason, some companies are initially concentrating on one-act plays.

There are also contractual problems. Successful plays or musicals are snapped up by Hollywood, who are still able to outbid cable producers. The going rate for a playwright whose work is turned into a movie ranges from £12,000 to £500,000. A cable company might have to pay up to £150,000 for licensing rights for a television relay. At this price most companies can only afford licensing rights. Some, however, notably Home Box Office, actually buy all the rights to stage productions in which they are involved. They thus acquire the residual rights to sell the theatrical property to other outlets in the United States or throughout the world – in theatres, on film, video cassette, video disc, satellites and on broadcast television. What may be a flop in one form of presentation may be a hit in another – at home or abroad.

As is evident from the Covent Garden project in the United Kingdom, there is a growing awareness that a good deal of money may be made through the sale of rights to the new television markets throughout the world. The emerging television production companies, such as Home Box Office, and the major film companies will be approached more and more by theatre producers who want to do a deal about selling residual rights. Equally, entrepreneurs, financiers and sponsors will be alerted to the possibilities of new kinds of investment.

The arguments advanced for the exploitation of stage production in the interests of the theatres themselves and as a means of meeting the increasing demand for television programmes, can be equally applied to the exploitation of other cultural activities to the same ends: art galleries and museums, for example, when they hold special exhibitions. More invention will be required by the television producers but the finished programme, produced in association with a book, will help interest people in the work of the gallery or museum and help affray costs. Add the public performance of music and the door is opened on a new world for many television viewers. It is the kind of programming developed and fostered by the public service broadcasters in Western Europe. It is now extending to America and elsewhere, but it is the European broadcasters who should be exporting their culture in television form. It should not be exploited by others. The means of funding productions is now established: Covent Garden is the example.

SPORT

Sport has universal appeal. The objectives are simple: hitting a ball into a hole or over a net, or winning a race. There is conflict between teams, as in soccer, or between individual contestants, reaching gladiatorial heights at times, as in the Men's Singles Final at Wimbledon. And there is always an end result.

In terms of television coverage, sport costs less per hour than all other programmes. No scenery is required, no rehearsal is possible and the viewer is provided with a ringside seat. In the BBC, for example, sport is almost eight times cheaper to produce than drama programmes and represents about fifteen per cent of all output on BBC-1 and BBC-2. Instructional programmes complement the coverage and help viewers to appreciate the finer points of a particular sport, or to learn how to do it themselves. Television probably helps attendance at minor sports (e.g. show jumping, gymnastics and skiing) and has no effect on sell-outs

(Olympic Games, cricket test matches and Wimbledon tennis), but may affect attendances for average football, cricket and athletic contests.

In the United Kingdom most sports clubs face continual inflation, rising costs, fewer opportunities for increases in productivity, and market pressures against further increases in gate money or admission fees. Soccer clubs are a good example. Annual attendances have nearly halved in the last thirty years, from forty-one to twenty-two million, and the debts of the top twelve clubs reach around £25 million. Soccer can no longer rely on a captive audience. Supporters have been driven away by high admission prices, poor accommodation, indifferent catering, and latterly hooliganism. In 1980 Southampton, one of the top clubs, had gross receipts of £2.1 million, of which ten per cent came from commercial activities, advertisements and sponsorship; yet the club suffered an overall loss of £954,000.

In February 1982 soccer club chairmen met at Solihull, near Birmingham, to plan a strategy for the future. Their plans include a £5 million advertising campaign to attract the seven and a half million stay-at-home television supporters back to the live game; and more sponsorship to help finance matches, including a £2.5 million deal with the National Dairy Council.

Cricket, the other British national game, was facing an even gloomier future as far back as the fifties. In 1963, however, encouraged by BBC Television, they introduced one-day Sunday matches which brought new vigour to the game. At the same time they adopted a more commercial attitude towards exploiting the game and club facilities. Some clubs now get up to half their income from squash courts, indoor cricket schools, bars, lotteries, advertising and sponsorship. Every major national competition is now sponsored.

The amount of money spent annually on sport in Britain by sponsors is estimated at £50 million, with 700 companies and fifty sports and leisure pursuits involved. In 1981, of the 1900 hours of television sport 593 were of sponsored sports, nearly a quarter supported by the tobacco companies. There is pressure on the government to ban sponsorship of sport by tobacco companies – tying in with the ban on cigarette commercials. In 1981, Embassy cigarettes, who sponsor the World Snooker Championships, had seventy-three hours of coverage on BBC Television and achieved a fifty-three per cent audience share. So far the government has refused to restrict sponsorship, though it has concluded two voluntary agreements with the tobacco industry which allow the amount of money spent on prize money to rise from £4.5 million to £6 million. In return, the industry is to invest £3 million in independent medical research. All advertisements for tobacco-sponsored events will, in future, have to carry a health warning.

In addition to encouraging sponsorship of their sport, some British Amateur Associations are permitting their members to advertise commercial products and services and still retain amateur status. One of the first to do so was the Amateur Athletics Association whose top runner, Sebastian Coe, was the first to be seen in commercial advertisements on television. The Association takes a percentage of his earnings to invest in the future of the sport, but it will still leave him a rich man. Amateur sportsmen of his stature will become attractive propositions to independent sports promoters. Coe is already signed up by Mark McCormack's International Management Group (IMG), a £100 million worldwide company with its head office in Cleveland, Ohio. Among other things, IMG markets the Nobel Prize awards, Miss World and the Pope's visit to Britain. It also represents 250 world-famous sportsmen and women. It produces television shows and promotes sports events, and owns what is claimed to be the world's largest sports-film company, Trans World International.

Entrepreneurs in the future will not only promote events but will also secure television rights for sale throughout the world. This was well illustrated in 1981 by the Golf Tournament at Sun City, South Africa's new £50 million resort complex. Five of the world's top golfers and ten 'showbiz' celebrities played in what was described as the century's richest tournament. The first prize money was £500,000. On top of this the last five holes were sponsored by South African companies for around £25,000 each. These were the television holes. SABC, South African Broadcasting Corporation, which did the filming, were given in return free rights for television in South Africa, while the organisers sold the international rights in Britain, West Germany, Japan and America. The TV rights were important because Sun City could not have recovered the million dollar prize money or the other expenses from admission charges alone.

Sports promotions of this kind will develop further, in line with the growth of new television markets, as can be seen already in American domestic television. The American networks have carried about 1400 hours of sport for the past few years, mainly on weekend afternoons. Their affiliates have made it known that they do not want more network sports. Cable affiliates, on the other hand, want the cable networks to fill all of their time, ideally twenty-four hours a day.

ESPN, Entertainment and Sports Programming Network, owned by Getty Oil, does just that, twenty-four hours a day, seven days a week, with almost every kind of sport, amounting to some 9000 hours a year. It started in 1979 and is now distributed by satellite to eleven million subscribers, ten per cent penetration of American TV households. Their forecast is thirty million by the middle of the decade. When ESPN reaches fifteen per cent penetration they will become eligible for metering by Nielson, a key audience research company, on the same footing as the broadcast networks. ESPN then becomes significant in terms of advertisers.

USA, another cable network that carries a good deal of sport, is owned in part by Madison Square Garden and ties in with sports activities there. It is scheduled ten to twelve hours a day, seven days a week and is distributed by satellite to over eight million subscribers. Three independent television stations – WGN-TV Chicago, WDR TV New York, and WTBS Atlanta – are also distributed by satellite. Known as superstations, they are on the air daily throughout the twenty-four hours, and rely on regular sports coverage.

The cable networks and the superstations are funded by advertising. In 1981 one corporation, for example, contracted to spend £11 million for a ten-year package with ESPN. In some cases costs are recouped by a charge on the cable company, based on a few cents per subscriber. Copyright problems are beginning to emerge over the relaying of games outside their 'territory'. The sports clubs themselves want a rake-off and there are legal proceedings pending.

Another incursion into television sport is STV, over-the-air subscription television. STV stations exploited local baseball games early on. In 1979 ONTV in Los Angeles signed a three-year licence with the Dodgers amounting to almost £400,000. Then following their success with the first of the Sugar Ray Leonard championship fights, which netted the station £500,000, they raised the subscription charge from $10 to $15 for the Leonard-Hearns fight in September 1981. Half of ONTV's 383,160 subscribers paid up, producing a total of $2.8 million. STV stations in three other cities also carried the fight and netted an additional $1.4 million. The fight was also taken on pay-tv by twenty-eight cable systems, including the two-way interactive QUBE in Columbus, Ohio. Viewers scored the bout round by round and the majority had Hearns ahead nine rounds to four when Leonard knocked him out in the fourteenth round. Overall the fight was reported to have netted the equivalent of £4.5 million from STV and pay-tv subscribers.

STV successes have prompted pay-cable companies to look again at the money-spinning potential of televised sport, especially as new 'addressable cable' becomes more widely available, making pay-as-you-view services practicable. Home Box Office already carry about sixty-five to seventy events a year as part of their general pay-cable service, but their emphasis is mainly on Hollywood films.

ABC Video Enterprises, a wholly-owned subsidiary of ABC Television, is already in the cultural ARTS cable network with Warner Amex. During 1982 they are planning to start a four-hours-a-day Women's Network with Hearst newspapers, and two satellite news channels with Westinghouse. They also intend to launch the first national pay-tv sports network with Getty Oil, starting with one major event a month. This will be presented in prime time on the existing ESPN service, for an additional fee yet to be determined. Subscriptions will be on an annual or semi-annual basis and will be offered to existing ESPN subscribers. The service will also be made available to STV stations. If successful, it will spin off a separate pay sports channel. Boxing will play a big part, and so will tennis 'specials' like the Borg–McEnroe challenge match in 1981.

Professional football still attracts prime-time audiences on network television at fees in the order of £120 million a year. The clubs are interested in pay-tv but not as pioneers. Though baseball is well established locally on cable television, the organisers of the World Series say they need the network exposure to promote the game. Basketball is also very much a local television event. Special baseball and basketball matches might be the first to be transferred to pay-tv at the local level. The 'Seattle Supersonics' basketball team have, in fact, taken the initiative and leased a cable channel which they intend to operate themselves on a pay basis, instead of selling rights to cable operators and programme companies, thus eliminating the middle man. The five-year contract cost over £500,000. All ninety pre-season and regular season games will be available to subscribers for around £60 through seven different cable systems in the area, with a potential of 216,000 subscribers. There are plans to extend the service to include high school games and a sports results service. The channel capacity varies from ten to thirty channels, so other sports such as baseball, football and hockey could be accommodated. In the United Kingdom there has been talk of similar city services based on coverage of local sport and news. These would be low-power broadcast transmissions funded on a subscription basis, similar to the ONTV subscription station in America.

For the new ESPN pay service finding a major national event a month is not going to be easy. For this reason the USA cable network have no plans to go into pay-tv, and Home Box Office has no intention of changing its current coverage. It is more likely that closed-circuit television in theatres will go in with pay-tv and share sporting events. 'Teletrack' is the name of a closed-circuit system in New Haven, Connecticut. A small theatre shows 'live' New York sports on a big television screen. The entrance charge is $5, and the theatre also houses a betting shop. For big fights you pay an extra $5.20.

The idea of sharing is developing elsewhere. CBS Sports owns the broadcast rights to the US Masters Golf, which it packages for the CBS network. In addition the company also covers the big daytime events at length for the USA network. It is this kind of specialist service that will also attract pay-tv interests.

Whatever the combinations, interest is not going to stop at exclusively American sports. ABC television paid £120 million for the American television rights for the 1984 Olympic Games. It has agreed to share coverage with ESPN and some events will possibly be shown on the pay service exclusively to recoup costs.

There are other sporting events that will attract the attention of pay-tv and closed-circuit television companies in America and elsewhere. The kind of events, for example, listed in the BBC's announcement of new sports contracts in 1982: the FA Cup Final, Wimbledon, the Grand National, athletics from Athens, Moscow and Brisbane, Test Cricket, the Rugby International Championship, the British Open Golf Tournament, the World Snooker Championship and the World Darts Championship, and all overseas Grand Prix. Traditionally, contracts of this kind have been basically for domestic transmission. The question of other rights for distribution by cable, satellite, pay-tv and home video, has been of secondary interest to the production departments within broadcasting organisations. The field has been left open for others to exploit.

European broadcasting authorities should recognise the threat to their operations if bigger interests with a greater purchasing power like American pay-tv enter the field. They could be relegated to the role of SABC in coverage of the Sun City Golfing Tournament. Equally, sports organisations could lose their independence and become slaves of the television industry.

There is still time in Europe for sporting bodies and the broadcasting organisations to work out partnerships that cover all aspects of television coverage to their mutual benefit. The 'Covent Garden' solution in the theatre could serve as an example. The broadcasters supply the skills and resources for the television coverage, the sporting bodies arrange the events, and third parties provide the funding, as private investment or as sponsorship or both. The initiative could come from the broadcasters or from individual sporting organisations. The important fact is that European television authorities should be involved, not foreign companies. This will not only protect domestic coverage but will maintain standards of European broadcasting and at the same time ensure that profits feed back to the sports organisations, not to the pockets of international entrepreneurs.

NEWSPAPERS

Newspapers and television have much in common. Newspapers have the writers, reporters and editors who understand rapid information processing. Television employs the same kind of journalists in its news and sports programmes. Both carry a great deal of feature material, though newspapers carry more articles of specialist interest than television can afford in prime time. Both share the same news agencies. These offer 'wholesale' teleprinter services from which newspaper and television editors select 'stories' for their separate 'retail' outlets.

Newspapers and television also share similar editorial problems in the interpretation of 'news values' and in the preservation of editorial independence. Usually they are legally constrained by the same laws, and have to exercise the same kind of judgement in matters of taste and decency. Television has the greater problem in that it brings the reality of a situation into the sitting-room, as opposed to a third-party report in a newspaper. Television intrusion into private grief, for instance, can be misconstrued as crude and insensitive compared with a newspaper feature on the same subject.

Television audiences and newspaper readerships obviously overlap. In the United Kingdom almost every adult watches television and more than thirty million people read a daily national paper. In cost terms the consumer pays about the same price per day for a tabloid popular, 15p, as he does for a day's output of all BBC Radio and Television services, and twice as much for the specialist *Financial Times*. Newspapers, though less immediate, exist in permanent form and can be scanned and read at any time, in any place, and in whatever page order desired.

Newspapers nonetheless face a bleak financial future. The problem worldwide is one of too many titles chasing too little revenue. In Britain, one forecast is that three out of the nine daily national newspapers published in 1982 will have closed down by 1985. British productivity is under half that of Germany and the United States, and delays in adopting new production techniques, type-setting by computers, for example, have prevented cuts in manning levels. According to an independent report in 1981, serious overmanning has been apparent for many decades, with the unions in effect acting as labour contractors.

Only regional newspapers continue to thrive. They operate on lower production costs, on an assured income from classified advertisements and local retail trade. Some newspaper groups, the Westminster Press, for instance, have launched free newspapers supported totally by revenue from advertising. The *Standard Recorder* reaches 177,000 homes in south-east Essex and replaced the *Standard* and the *Recorder* which had a combined circulation of only 40,000. The new paper is thirty per cent editorial, with six editions, offering advertising flexibility.

In Germany, despite better productivity and the introduction of new technology, serious papers like *Die Welt* are suffering losses in the order of £8 million a year. In the United States the closure in 1981 of the *Washington Star* after 128 years illustrates the dilemma facing newspaper owners. The metropolitan area of Washington has three million people, in good jobs with high and steady incomes. For many years it supported two newspapers, the *Washington Post* in the morning and the *Washington Star* in the evening. Whereas morning papers have remained profitable, evening papers in most big cities have become losers. Chicago has no evening paper; London now has one in place of three. One of the problems facing evening papers, of course, is maintaining speedy delivery in rush hour traffic, which does not apply to very early delivery of morning papers. Moreover, breakfast television provides less of an alternative to morning papers than early evening programmes do for the tired worker home from the office or factory.

The impact of television on newspapers has been considerable: the obvious competition; the siphoning of advertising revenue and newspapers' growing dependence on television for news. For many years the popular press has been devoting increasingly more space to television. Newspapers now look to television more and more to initiate public debate on major social, economic and political issues. Increasingly, also, front-page stories in both serious and popular papers follow up television coverage of news events. Politicians are particularly aware of this. By timing their public statements to late afternoon, they ensure full radio and television reports during the evening, with follow-up reaction in the next morning's papers and on breakfast radio and television.

It is understandable, therefore, why newspaper proprietors have been investing in television rather than continuing to prop up failing newspapers. In 1981 the *New York Times*, in a £50 million deal, acquired cable operations of two companies serving Southern New Jersey, and the *Chicago Tribune* also bought up a cable system in New Jersey. In Germany, the Federation of German Newspaper Publishers may take a share in one of Europe's direct-broadcasting satellite projects based in Luxembourg. The question is whether they will go further and adapt their skills and resources to compete with television in providing programmes and services for the new television markets – cable, satellite, pay-tv and home video. There is little evidence to date that newspapers will try to emulate in picture terms what television broadcasting does well in the coverage of news and sport on a day-by-day basis. Newspapers, though, can exercise a different editorial eyeline, on a different time-scale. In 1980 in London the *Sunday Times* did precisely this when it co-operated with ITN, the Independent Television News Company, sharing their different skills and resources to produce a 'Review of the Year' for sale on video cassette.

More opportunities are likely to arise in the development of specialist information services. These are the areas where news agencies and newspapers are expanding. They are also the areas where national broadcasting can only make minor contributions. Some local television and cable stations in America do provide specialist services. A good example is KWHY in Los Angeles, which broadcasts thirty-five hours a week of financial reports from a studio looking like a Stock Exchange. They rely on news agency services and their own analysts. They also broadcast daily programmes in Japanese, Chinese and Korean. Both financial and ethnic services are supported by advertising. During evening hours they run an over-the-air pay-tv service of first-run feature films, Las Vegas-type revues and major sporting events.

News agencies, as the 'wholesalers' in the news business, are more active than actual newspapers in finding new distribution outlets for their services. In America all the main agencies – Associated Press, United Press International, Reuters, Dow Jones and Eastern Broadcast-

A video cassette, *The Year 1980*, coproduced in London by the *Sunday Times* and ITN.

ing Services – are actively involved in television. Each presents a 24-hours-a-day news text service, distributed to subscribers by satellite and cable or by telephone lines. A typical screen display consists of seven or eight lines of thirty characters. Services are built on time modules, from ten minutes to an hour. Subject-matter consists of national and international news, sport, business and financial news, and weather. Reuters also carries a daily horoscope, a sports quiz, a feature on Hollywood and advice on pet care. Charges vary from ½ cent to 6 cents per subscriber. The biggest system is Reuters with three and a half million subscribers. One forecast is that the annual income for electronic news text service will grow to £250 million by 1985, divided evenly between consumer and business interests.

Small town American newspapers also produce electronic news services for local cable distribution. First in the field was the Ottumwa *Courier* in 1977. Two years later the *Globe Gazette* in Mason City followed suit. The paper now has 6500 cable subscribers. It has a capacity of 120 pages, seventy per cent news: thirty per cent advertising. Altogether there are more than twenty newspapers employing cable in their electronic publishing. Some of them are tying in with the agency news text services by taking a satellite feed. A satellite dish receiver costs about £10,000 over and above the £10,000 for the electronic equipment to generate the characters for the screen.

Broadcast teletext offers few opportunities for newspapers. Serious papers, serving specialist interests, will make better use of the two-way viewdata systems where up-to-the-minute information can be called up by telephone from a central computer, for display on the television set or printed out, and the consumer can automatically be billed as with telephone calls. In the United Kingdom providers of such services on Prestel include Reuters, Westminster Press and the *Financial Times*. In America Associated Press has been involved in a series of similar projects whereby personal computer owners within each newspaper circulation area can access a news data base for about £2.50 an hour.

Newspapers are looking to television as a cheaper system of distribution. In Britain, on top of high production costs, national newspapers face increasing costs in physically distributing their products through wholesalers, retailers, news-stands or newsboys actually delivering papers through the letter-box. Already some papers distribute page by page electronically, from London to Manchester, for example, for local printing and distribution. The *Financial Times* does the same from London to Frankfurt for its European market. The *New York Herald Tribune* goes further and sends its Paris edition by satellite to Hong Kong for its Far Eastern market.

The United States, because of its size and regional differences, does not have a national press like Britain's. Only the *Wall Street Journal* and the *New York Times* approximate to national newspapers, and each has a specialised readership. In late 1982, *USA Today* will be launched by satellite from Washington to thirty-eight cities for local printing and distribution. This will be a national newspaper aimed at the mass market.

It is not known yet whether it will actually be cheaper to distribute newspapers in one form or another electronically, direct into people's homes for display section by section on the television screen or for printing out overnight, in colour if required, as developed in Japan. Paper costs will equate, but other uses for home printers will have to be developed side by side to make their installation economic. But television technology will help reduce newspaper production costs. When computer typesetting is widely established, a journalist anywhere in the world will be able to type his story on to punched tape or audio tape and reproduce this over the telephone to his paper, where it will be automatically typeset. The output from new electronic still cameras will also be played over the phone to provide the pictures to back up the story. If the timing were right, a journalist for the *New York Herald Tribune* in Hong Kong could see his own story on sale on the streets within the hour.

Newspapers are unlikely to be serious rivals to established broadcasters in the provision of programmes or services in the traditional areas. There may, however, be conflict between the broadcasters and agencies and newspapers over who should have the rights to transmit teletext. In Germany the Newspaper Proprietors Association and the German broadcasting organisations have become bitter rivals over the question. It is part of the survival fight. Television and newspapers are inevitably in conflict for advertising revenue. This will come to a head in Britain in 1982 when the introduction of Channel Four will, in its first year alone, be taking some millions of advertising revenue from Fleet Street newspapers. It is a pattern that is being repeated round the world.

MAGAZINES

Like newspapers, magazines have much in common with television. They employ the same kind of writers and face similar editorial and legal problems. They also cover the same subjects, but whereas a newspaper has to serve a variety of interests like a television schedule, a magazine is like a television programme or newspaper feature in that it deals with one subject area only, on a continuing weekly or monthly basis. Thus magazines can treat subjects in greater depth than television and newspapers, and appeal more to specialist readers who are willing to pay to satisfy their minority interests. For comparison, magazines can be conveniently grouped under three headings: News, General and Specialist.

In America news magazines flourish for the simple reason, so it is argued, that no newspapers regularly reflect the affairs of the nation in depth like serious newspapers in Britain. Thus *Newsweek* and *Time* magazine appeal to readers throughout the States who need to keep themselves informed about national and international affairs. And the information has to be in permanent printed form so it can travel and be read or referred to, as and when required. This, of course, is not possible with television which tends to be concerned more with news on a day-to-day basis and which is its strength. *Newsweek* and *Time* magazine, being published in English, have also been able to develop an international market of people who wish to keep themselves informed and who will pay for the service. In Western Europe news magazines also flourish, *L'Express* in France and *Der Spiegel* in Germany, for the same reason – the absence of national newspapers on the British pattern. Furthermore, continental Sunday newspapers have not yet developed the idea of colour magazine supplements as they have in the United Kingdom. These have successfully attracted an increasing proportion of the available advertising revenue at the expense of news magazines.

The magazine *Now!*, for example, was launched in September 1979 as the British counterpart to *Time* and *Newsweek*, *Der Spiegel* and *L'Express*. It sold 380,000 copies a week. After six months sales had halved and in April 1981 it closed down. The crucial issue was not, in fact, circulation but advertising revenue. *Now!* magazine was in competition with its nearest rivals, the Sunday newspaper free colour supplements, not so much editorially, but in fighting for its share of the limited advertising revenue. In 1982 there were five national Sunday colour supplements, the *Sunday Times, Sunday Telegraph, Observer, Sunday Express* and *News of the World*. Each carries about two-thirds advertising and more or less competes for the same accounts. Size varies from 30–120 pages.

The news magazines that survive in Britain are those that concentrate on one or other aspect of the current scene: *Time Out* on entertainment; *New Scientist* on science and politics; *New Society* on sociology. *Private Eye* is the exception; it cannot be tied to any particular market, yet it sustains a fortnightly circulation of 170,000 and carries very little advertising. The traditional political–literary weeklies do not fare so well. In the past twenty years, the *New Statesman* and the *Spectator* saw their circulations halved. Only the *Economist* with its specialist business interests has increased its circulation threefold over the same period. In terms of exploiting television, news magazines have the same opportunities as newspapers to diversify their activities into teletext, viewdata and cable systems.

Television, by its nature, affords more opportunities for general magazines which are concerned with the big areas of public interest: Do It Yourself, Homes and Gardens, women's interests, and the broad area of leisure pursuits as distinct from professional or intellectual interests. There is overlap with television in subject areas, but television broadcasting cannot compete in depth on a regular basis with the wide range of interests of general magazines, although the new television markets will be able to do so. For this reason, far-sighted magazine publishers are turning their attention to cable, pay-tv, satellites and home video. For some the interest at this stage is limited to investing in order to get a foot in the door. Others have been more adventurous and set about translating their skills and resources to television.

The first to do so in America was the *Family Circle* magazine owned by the *New York Times*. In partnership with cable television they produced a series of half-hour programmes based on the magazine's special feature 'Great Ideas' about a whole range of women's interests. The series started on SPN Cable in March 1981, with five programmes a week, screened twice daily. In addition to six minutes of advertising a day, the series also attracted an initial sponsorship of £3500 per week. The series has since moved to CBN Cable and extended its potential subscribers from 2.3 million to over thirteen million.

The publishers of another general magazine, *Better Homes and Gardens*, have coproduced thirteen half-hour programmes, 'Better Homes and Gardens Idea Notebook' with an independent production company. Each programme is shown six times a week at varying times on the USA Network. The series is supported by advertising. The idea is not just to take a magazine article and translate it to television, but to draw on the magazine's information and staff resources in order to create television programmes.

The first involvement of Time Inc., was a sixty-minute sports feature, for Home Box Office in May 1981, on a pay-tv basis. *Money Matters* later in the year was another one-hour programme for the HBO network. It was produced in close co-operation with their *Money* magazine, with the staff taking part. The production cost £125,000 and no direct advertising of the magazine took place on the programme, but the hope is that sales will increase.

Ladies Home Journal, Redbook and *Ms* magazine are each involved in developing women's programmes for cable networks. The Hearst Corporation plan to draw on the expertise of their editors and writers of *House Beautiful, Good Housekeeping, Harpers Bazaar, Cosmopolitan* and *Town and Country*, in the development of television programmes.

The Playboy Channel started in early 1982 with an estimated seven and a half million households. It takes over the Escapade pay-tv channel, which carried X-rated films between 8 pm and 6 am seven days a week. The Playboy Channel say they will feature 'sophisticated adult programming' based on the successful editorial departments of the magazine. Penthouse Entertainment Television, otherwise known as the PET Network, also plan to start in 1982 for twelve hours a day, from 8 pm to 8 am. The Penthouse Channel will be aimed at the 25–45-year-old market, while the magazine will continue its appeal for the 18–25-year-olds. They hope to reach one million cable households. The launch is reputed to cost £2.5 million, put up by Penthouse and their partner Telemine. Editorially they plan to draw on the four related magazines, *Penthouse, Forum, Omni* and *Variations*, to provide what they call 'a format of the future, science fiction and sex'.

The proliferation of cable channels generally has created its own problems for the cable operators and magazine publishers on how to keep viewers informed of programme plans. There are at least fifteen different cable guides, ranging in circulation from 15,000 to two and a half million. Some are given away, others are sold. This is not yet a problem in Western Europe, and the limited services available are accommodated in one or other magazine. In fact, the biggest-selling weekly general magazines in Britain are *Radio Times* and *TV Times*. Programme guides will adapt easily to teletext transmission and, in the long term, to printing in the home. *Radio Times* and *TV Times* will be cautious, understandably, because their present financial buoyancy results from their being a magazine in constant use seven days a week and thereby a boon to advertisers.

Cable can be successfully exploited for subjects of interest to audiences measured in 100,000s. Video cassettes and video discs are a better proposition for dealing with specialist or minority interests. An early pioneer in the United Kingdom was IPC, the International Publishing Corporation, which publishes a variety of general and specialist magazines. In 1978 they bought up existing films on swimming, tennis, fishing, motor racing and others that tied in with one or more of their specialist magazines. They had the films transferred to video cassettes, nicely packaged, and marketed them through their respective magazines, selling at around £40.

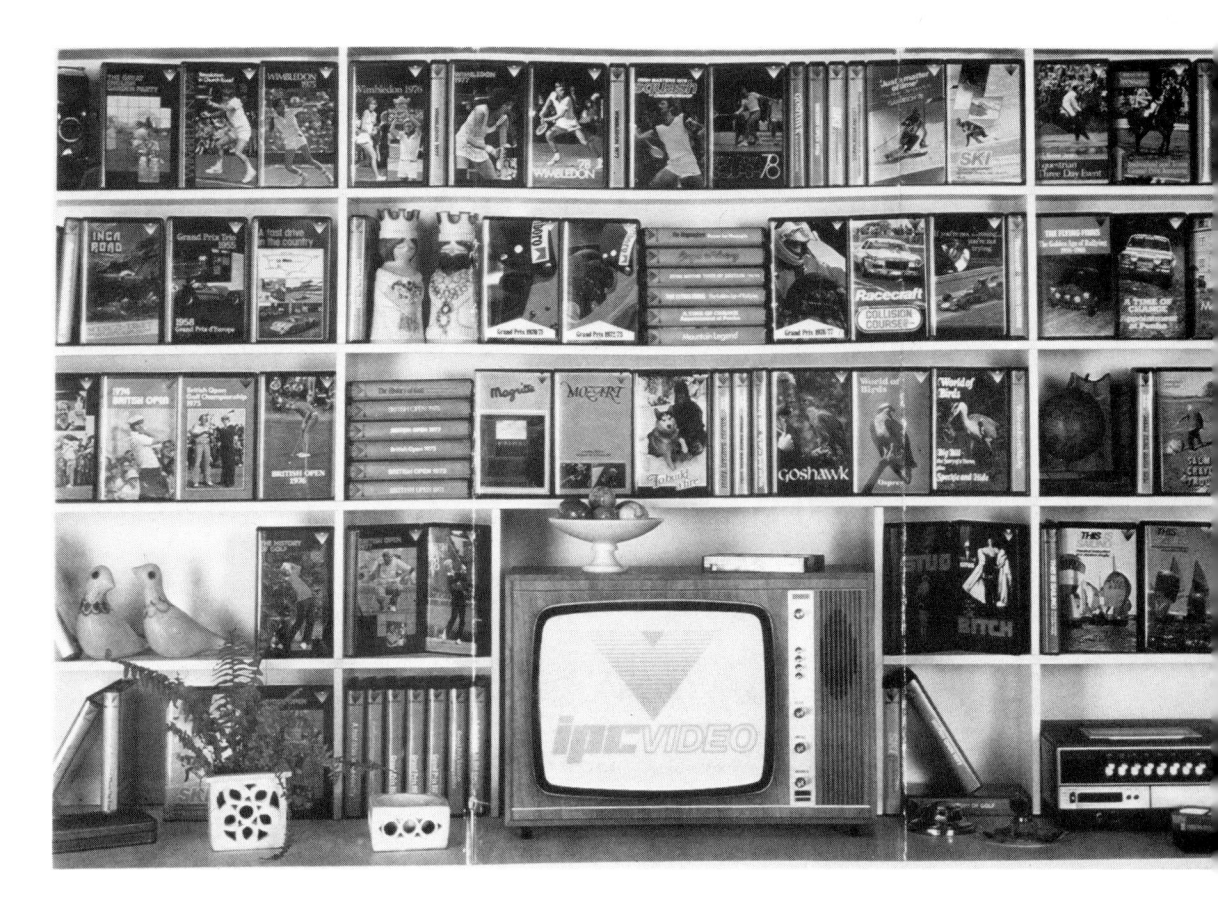

Of all the specialist areas, sports magazines are most likely to lead the way into television. Sport is concerned with action of one kind or another which makes for good television. Moreover, sport is usually associated with clubs and they, rather than the individual members, will be able to afford video-cassette machines and the prerecorded video cassettes. Some golf clubs have gone further and purchased an electronic camera which a member can use with a video-cassette machine to record his play for subsequent playback and discussion of his performance.

What television can do is to show how things work, how parts interrelate in action, how to paint, how to do...anything. The printed word complements the action. By 1990 some magazine publishers will have become television production houses, turning out recorded 'video magazines' for distribution to local cable and television stations and to the home video market, along with their printed magazines. As with the record industry, the magazine industry will adapt its distribution network to handle the sale of video cassettes and video discs alongside their associated magazines. This will ensure a common point of sale in the

An IPC brochure showing their range of general and specialist-interest video cassettes, particularly sports programmes that tie in with their specialist magazines.

High Street shop or department store. Moreover, specialist magazines are usually associated with club activities and cassettes and discs selling at perhaps £10–£15 could be distributed on a mail order basis.

Specialist magazines will also make use of teletext. In America there are already comprehensive viewdata-type services for farmers: 'Agvision', sponsored by a weed-killer company; 'Instant Up Date', which charges £50 a month for the hire of the home terminal and unlimited information retrieval; and 'Project Green Thumb', based on weather services, farm prices and so on, working with the local state agricultural service. As teletext and viewdata services develop in America and in Europe, there will be opportunities for magazines to exploit this potentially profitable relationship between electronic printing and magazine publishing.

Magazines which specialise in puzzles and games, at one end of the scale, and home computers, at the other, will find new outlets in the manufacture of video games and computer programmes for use with two-way cable stations and with interactive video discs. There are five such cable projects in the United States, called 'Play Cable'. About a dozen video games are available for a subscription of £5 a month. The home printer will be an important factor in printing out video games or computer information, official coupons and competition entry forms.

Experience already shows that magazine publishers can translate their skills and resources to the production of television programmes and services; in general interest areas to cable and pay-tv; for specialist and minority interests to home video, teletext and viewdata. More opportunities will arise as the new markets expand.

BOOKS

Book publishing and television broadcasting share the same aims overall: to entertain, to inform and to educate. The balance between the three varies within different publishing companies and television networks according to whether, for example, they are publicly or commercially funded, and the degree of public responsibility they accept or which is placed upon them by governments or their agencies.

They also share common subject areas. Television schedules match publishers' catalogues and range as widely, from the Bible to the cheapest paperback. They exercise the same kind of editorial judgements in selection and in matters of taste, and share similar legal constraints. They also share many of the same writers. On the other hand, printing technology is vastly different from television, though there is greater compatibility as electronics permeate printing techniques. The time-scale of publishing is longer than television, but the end product has substance and can be handled. It also has permanence, unlike the transitory nature of the television screen. Books have to be physically distributed through wholesalers to retail outlets. There has so far been no attempt to distribute books electronically like newspapers.

Book publishing the world over has problems. These arise mainly from the effects of inflation and the general world recession. Printing costs in Britain in 1981 were so high that an increasing amount was printed overseas. Although doubling administrative problems in proof correcting and in control of standards, this still saved the publishers money. The cuts in public spending in 1981 severely reduced the number of new books going into education and public libraries. New titles published have dropped from 48,000 in 1980 to an expected 38,000 for 1982, and consumer spending on books has fallen by ten per cent. In the United States, although book sales increased by eleven per cent in 1980, the growth represented higher book prices rather than greater volume, and American publishers continue to complain about a slump in sales and profits.

Another major problem is the combating of piracy. Book piracy is now a booming business. Photo-copying is common practice. A recent survey of ninety-seven schools in Scotland revealed that 66,428 copies of literary and musical works were made in six weeks. It is estimated that the British book trade alone is losing over £1 million per year through the activities of book pirates. The Publishers Association, with the support of the Association of American Publishers, is involved in legal action against book pirates in Singapore and India. In a sense, piracy is proof of the worldwide interest in English-language publishing. The British Publishers Association see new markets growing in the USSR and China and in the developing countries, despite the effects of piracy.

The Association also takes an optimistic view of developments in other media: they see no reason, for example, to believe that the introduction of the new Channel Four in Britain in 1982 will do anything other than stimulate interest in books, or that new forms of information provision such as 'electronic publishing' will do anything other than create new opportunities for publishers. The transfer of law books to a computer base is an indicator: 100 million words of English legal text are stored in Britain – the equivalent of 300 printed volumes – with access to 3000 million more words stored in America.

The question is how far publishers are going to invest in television. They have effective business deals with the feature film industry. Best-selling fictional works invariably end up as feature films. Television cannot match Hollywood. It copes best with non-fiction and there have been many tie-ups in the non-fiction area from which both books and television have benefited. In 1981 Carl Sagan's *Cosmos* reached the American best-seller list at the same time as the TV series of the same name was being shown on the PBS network. There are similar situations in Britain: Alistair Cooke's *America* and David Attenborough's *Life on Earth* are but two examples. In these situations the author stands to make

Some best-selling BBC books published in association with television series.

more money out of the book than the original television programme. Publishers are aware that advertising and good reviews are not enough to guarantee a book's success. Some kind of link with television is required, together with promotional interviews with the author on radio.

Television, however, has more to offer. Whereas a book, at best, analyses a situation, television shows things, and relates one to another in reality. The introduction of the Laservision video disc will enhance this ability, in being able to run to and fro in slow motion, to single out at random any of the 54,000 individual frames on each side of the disc, and to have two languages, or speech on one channel and sound effects on the other, or simply stereo sound. Books and television become complementary.

This was foreseen for home video as early as 1977 by Donald MacLean, who was in charge of videogram production at EMI in London. He had established an informal relationship with Collins, the publishers. Their most quoted project at the time was a video disc on birds, with an accompanying book. All the facilities of the disc would be exploited to show birds in action, with natural sound on one sound track and commentary on the other. There would be captions to guide the viewer to the appropriate parts of the book, which would include information, fine detailed drawings and maps (which television cannot effectively reproduce). The book and the disc could be more interrelated on a 'programmed' learning basis. For one reason or another the project never got off the ground, and in 1982 the BBC, drawing on its extensive Natural History Film Library, released a video cassette on British Birds. There is also a video-disc version, which has 170 teletext type captions complementing the spoken commentary, and has the natural sound on one track and the commentary on the other.

Mitchell Beazley, the British publisher, pioneered in 1969 a new style of consumer-orientated books which combined photographs, graphics, maps and text, often spread over two pages. One of their first successes was *The World Atlas of Wine* which has sold a million copies in many languages, the visual material being common to all editions. It is only one step further to produce a television version. Mitchell Beazley has long argued for a video-disc encyclopaedia that would exploit the 54,000 frame storage. Meanwhile, they are the first British publisher to have a series of television documentaries (on Africa) accepted by Channel Four. An important factor in this kind of development was the takeover of Mitchell Beazley by American Express in 1981. They will provide the basic funding; Warner Brothers, who are also financially involved with American Express, will have the means of production and distribution; and QUBE Television and others, owned by Warners, will provide television outlets. Moreover, Mitchell Beazley has also gone into partnership with Thorn EMI in London to produce and market 'multi-media' packages – selling a book on cookery, for example, with a related television programme. With a £3 million investment they plan eighteen projects over the next three years, using video discs.

In 1977 the American publishers, Time-Life, noted for their stylish coffee-table books, were moving into the production of coffee-table video cassettes. In the light of experience they have now invested in Home Box Office, the biggest American pay-tv company. This will guarantee their television outlets. These are indispensable if the television side of publishing is to be financially successful. Equally, television exposure is essential to launch an associated book. Thomsons, the British publishing house, invested something approaching £250,000 from their North Sea oil revenues on a definitive series of documentaries called *The Commanding Sea*, made by independent producers. Once completed they had to find television outlets for the programmes and to launch the projected book. In the end, the BBC bought the documentaries and copublished the book with Thomsons.

More and more publishers are recognising the complementary relationship of books and television. Just as books are republished in different versions for different markets, so they will be adapted for television. Popular subjects such as gardening, needlework, do-it-yourself and cooking are obvious first choices. They demand more than illustration, and need demonstration, which television can do well. In terms of distribution books, with their associated video cassettes or video discs, will be sold through existing outlets (sharing common advertising and points of sale) and mail order book clubs. This is already happening in Western Europe through W. H. Smith in Britain, Hachette in France and Elsevier-NDU in the Netherlands.

In terms of television production there may well be a degree of conflict. The pattern established by the BBC is that their books spin off from their programme output, not the other way round. Moreover, the different production processes to make the programme and the book are carried out by specialist staff. Only the writer may be common. The aim is to back-time the two operations so that the book and the television programme appear simultaneously, each benefiting from the other. For example, the book can be promoted along with the broadcast programme.

In future, publishing houses will begin to recognise ideas that are as much potential television material as they are books. There is evidence already in the fact that the disposal of television rights is frequently part of an author's contract. When this happens a publisher will be faced with the problem of going it alone and commissioning the television production, making use of independent production and facility houses; or going into partnership with a television broadcaster to coproduce the programme and to copublish the associated publications on a jointly funded basis. The advantages of the partnership scheme is that production quality in each media is safeguarded, the television exposure is guaranteed, and the book is promoted along with the programme. Afterwards the programme can be sold to the other markets throughout the world with the book.

There will be no difference of view between publishers and broadcasters about the sequence of distribution — broadcast television, cable, pay-tv and home video. Broadcasting gives the biggest television exposure and goes first. On the other hand, publishers and broadcasters may compete for writers and creative talent. But when a partnership works effectively the writer stands to gain considerable financial rewards from having his work 'published' not only in different book versions, but also in a variety of television outlets. The arguments are strong for coproduction and copublishing. They also reflect the complementary relationship of books and television in terms of the reader and the viewer, as well as the publisher and the broadcaster.

EDUCATION

In the broadest sense, all television is educational. It provides a continuing awareness of the world about us. Drama, documentaries, news and current affairs, even light entertainment, are part of this pattern. In most countries television also provides specific educational services directed to pupils in schools and to students in colleges and in their homes. These are in the tradition of radio broadcasting, and complement the use of books and films and other teaching aids in 'extending' the classroom or lecture hall. The nature of schools broadcasting is largely determined by the educational system it serves. In America, where education is the responsibility of individual states, schools broadcasting operates at a regional or local level. In the United Kingdom it operates mainly at a national level.

Broadcasting to schools in Britain began in 1924, two years after the start of broadcasting. Today ninety-two per cent of primary schools and eighty-three per cent of secondary schools make use of television broadcasts. Three-quarters of secondary schools now have video recorders. This gives flexibility in matching the broadcast schedule to the school timetable, and there is legal provision for such 'off air' recording. As well as regular morning and afternoon programmes for primary and secondary schools, there are pre-school programmes for the under fives: songs, story telling, puppetry, animation and films. Research suggests that as many as nine out of ten under fives regularly watch these programmes at home or in nursery schools.

There are also further educational services for use in colleges and institutes, but primarily for use in the home. The aim, in the official language, is to give the viewer a progressive approach to some skill or field of learning, vocational or recreational; in other words: learning a language, art appreciation, car maintenance, care of the aged, and so on. Many series, as with broadcasts for schools, are supported by textbooks, pamphlets, films and cassettes.

BBC and the ITV companies produce and fund the programmes, but they work closely with National and Regional Educational Councils. Although the broadcasting organisations retain final responsibility for what is transmitted, they would be reluctant to broadcast or publish associated material that did not have the support of the appropriate Council or Committee. It is a partnership that is also shared between the BBC and the Open University, which was created in 1970 to enable working adults who are over twenty-one to study for degrees or to achieve some shorter-term educational objective. It is the most elaborate educational project anywhere in the world, utilising radio and television as instructional tools. The Open University has also developed a system called 'Cyclops' which links the television set to the telephone. This allows its students and teachers, many miles apart, to be able to draw diagrams or graphs and write on their television screens by using a light-sensitive pen.

There are around 70,000 students studying one or more of the 130 courses in their own homes and attending occasional summer schools. Each year the BBC produces 300 new television programmes and 300 radio programmes. In 1981 Open University programmes occupied almost thirteen per cent of BBC television network time. The cost of both radio and television broadcasts was in the order of £10 million, which the Open University reimburses to the BBC. The broadcasts are supported by printed material, making up what is called the 'media package'. The demand overseas for Open University and British educational packages generally is so great that in some areas exports are becoming a profitable business.

Because the Open University is built around radio and television transmissions, its teaching is public rather than private, and the system is open to all. In fact, over and above the registered students, programmes are eavesdropped by audiences in the order of 250,000. It has been shown that, for each of the 70,000 students who watch a television programme, at least a dozen other people follow the course with the same degree of attention, but without actually 'doing' the course. If video and audio cassettes replaced broadcasting overnight the University would no longer be open and public, but would become a private institution. So broadcasting is central to the concept of the Open University.

This is, of course, only possible when broadcast transmitters are idle. Conflict for time between educational interests and general programming has already arisen between the BBC and the Open University. One solution proposed is to share the Open University transmissions with Channel Four when it starts in late 1982. A similar conflict of interest exists in BBC Radio, where schools programmes occupy one network mornings and afternoons during term time. Overnight transmissions would free time for general use. The idea is for schools to record their programmes for later playback, using automatic time switches. The

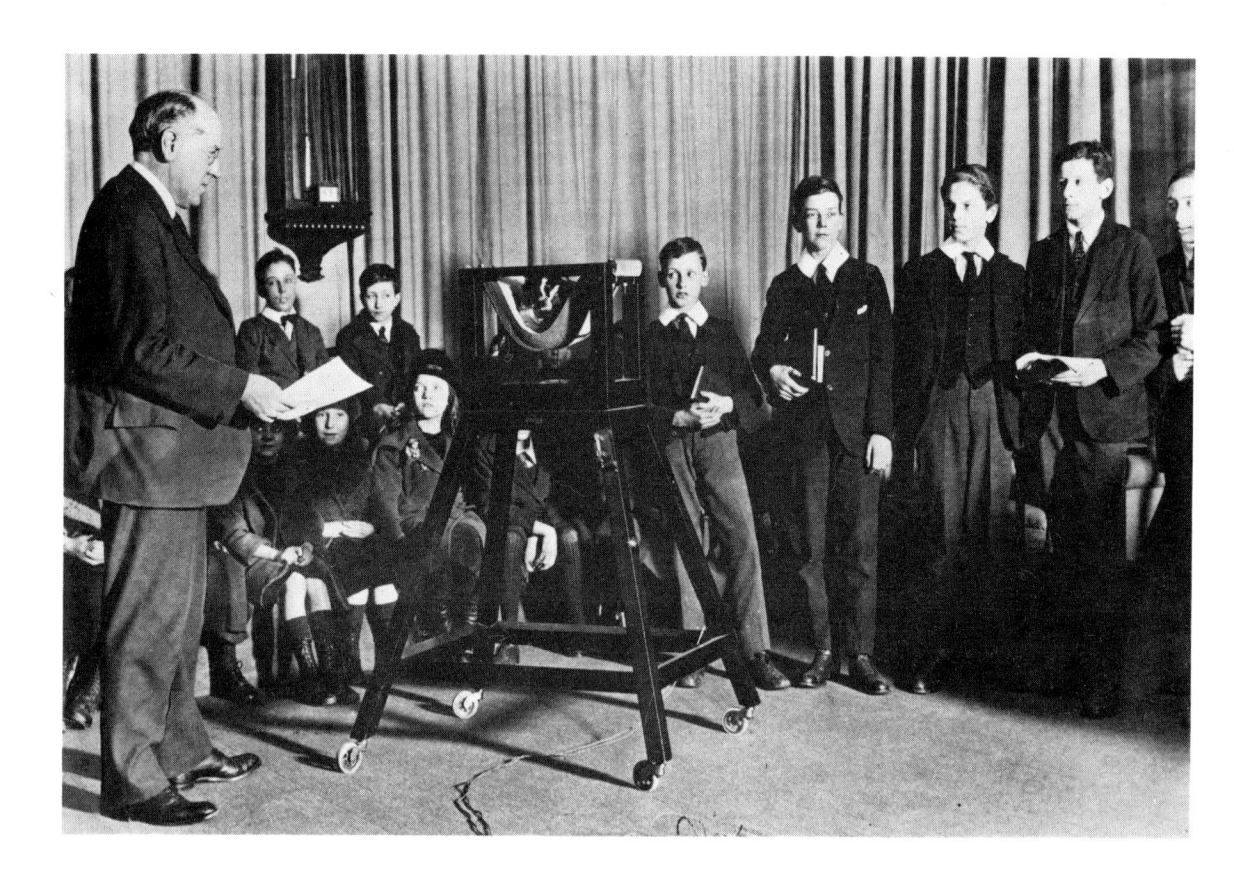

TELEVISION IN THE EIGHTIES

Left: the first BBC broadcast to schools in 1924; Sir Walford Davies gave a lecture on music to the Temple Church choir boys. Below left: a science lesson at a Hertfordshire school, 1982.

position may change dramatically when more distribution outlets are available, by satellites and possibly by cable.

With this in mind, it is important to consider educational television in America, where the number of television outlets is increasing. Traditionally, educational television in America was run at the local level with schools and colleges making use of films rather than broadcast programmes. This is continuing, but films are being transferred to video cassettes with considerable savings in costs. Granite School District, Salt Lake City, claims to be saving thousands of dollars every year and at the same time making the programmes more easily available. This is important since the school district comprises seventy-three buildings, and the Library contains 400 titles of elementary level, 800 of secondary level, and 1200 titles for the senior high school level. At university level in the States there are projects similar to the British Open University, but making use of video cassettes instead of broadcast transmission. In Tucson the University of Arizona has developed what they call a 'microcampus' programme. Students enrol and pay the same fees as other students. They do not attend college, but learn via television. Examinations and marking of work is handled through the mail. The objective is to provide a service for students and professionals remote from the campus. Something like forty-five one-hour cassettes form the basis of the equivalent of a typical three-credit college course. In the longer term interactive video discs will be available at less cost than the equivalent video cassettes.

Although American educational television originated locally, the development of satellite distribution systems has led to the idea of educational networks. The first began in the early 1970s, as a federal research project by the Appalachian Commission, covering the area from New York State down to Alabama. The commission built fifteen community dish aerials to receive satellite educational programmes originated in Washington, DC. The Appalachian Community Network is now carried by 145 cable systems, reaching 1.2 million homes spread throughout thirty-five states. Each system pays one cent per subscriber per month for the service. Subscribers pay tuition fees, about thirty per cent of which goes to the network. Other revenue is received in the form of grants and donations. At present no advertising is carried. In 1981 fifty colleges and universities subscribed to the network, in addition to individual households. Schedules range widely; there are sixty-four hours of programming a week, mostly in the further education field.

The American Educational Television Network came into operation in 1979 with the simple concept that professionals in many fields, by the nature of their work, often require continuing education. What better way to reach them, so it was argued, than through a medium which they could view in their own home. The Network is carried by 300 cable systems and twenty-two broadcast television stations and it claims a potential reach of over forty-seven million households. It is funded by grants and donations, and by sponsorship by national advertisers; some local advertising is available. It also charges tuition fees.

Other universities, notably Pennsylvania, have linked up stations over short distances by microwave transmitters. At present a four-hour block of programming, permitting up to eight course offerings, is repeated three times a day. Their intention, however, is to extend the service to twenty-four hours a day in six-hour blocks. In some systems a free toll number is available to provide for interactive learning or for student counselling. Interactive television is already exploited for local education on the two-way QUBE cable system in Ohio, and is being extended as two-way systems grow. In Cincinnati, for example, the 62-channel system will have nine educational channels, with South Western Ohio colleges producing the programmes from the PBS (Public Broadcasting Service) studios in Cincinnati. As in Britain there is legal provision for recording educational programmes.

When the PBS network was reorganised in 1980, three separate programme services were established: PTV-1 to develop prime time programming; PTV-2 to develop special interests; and PTV-3 for educational programming. By the end of 1981 PTV-3 expected 112 of the 170 PBS stations to carry its nine college courses, in conjunction with 500 colleges across the country. Students pay fees which go to the colleges participating in these so called 'tele-courses'. Local PBS stations and the network derive revenues in turn from the universities involved and from charges for supplying production skills and resources in the making of programmes.

As in general programming the television outlets for education in America proliferate. Unlike the United Kingdom there is no regulation either of the systems or of the programme schedules. Programme making depends on local academic initiative or commercial profit. Organisations such as the National Geographic, Learning Corporation of America and Time-Life have extensive libraries of educational programmes which are rented or sold on a commercial basis. Like educational textbooks, which are a money-spinner in the States, educational programmes can also be profitable because of the scale of the domestic market. *Sesame Street* is perhaps the best-known example. The United Federation of Teachers, among others, attempts a co-ordinated approach. In their weekly newspaper, which is read by 240,000 teachers in New York alone, they draw attention to scheduled programmes that could form the basis for discussion in the classroom. The Federation is also asked from time to time to endorse a pilot programme for a new commercial educational series to help its launch.

There is no direct government funding. Income is derived from subscriptions from viewers, advertising, sponsorship, university grants and donations (such as the recent offer by the publishing tycoon, Walter Annenberg, of the equivalent of £75 million to the Corporation of Public

Broadcasting to set up a National University of the Air, on the lines of the British Open University).

In the light of the American experience, what are the questions for educationalists and broadcasters in the United Kingdom and Western Europe? Will it be possible to sustain the kind of partnership developed in the United Kingdom over the past sixty years in Schools Broadcasting, Further Education and the Open University? In seeking answers it is important to distinguish between regulated programming made at the national level for broadcast, or for distribution by other systems to relieve the pressure on transmission time, and specialist programmes made locally for closed-circuit use within schools and colleges or in the home. There would seem to be no reason why national programmes should not continue to be produced in partnership by the broadcasters as now. Specialist and interactive programmes could be produced in universities, many of which already have the resources. Even schools or local educational authorities could develop programmes for their own use as electronic equipment become cheaper.

The supply of both general and specialist programmes will have to be matched to possible changing requirements of the classroom. Educational information systems will link television programming to computer-based teaching aids. It is significant that American Parent–Teacher Associations are already donating computers to schools, and in Britain the BBC has developed an educational computer to plug into the television set. In the long term each student will have his individual desk-top computer with its own screen, and with access to the school or college microprocessor and data banks, and through viewdata systems to area and national educational centres, involving satellites. The University of Surrey, near London, has already built its own communications satellite, which was successfully placed in orbit in October 1981.

Educational teletext is already available in Britain. Experience has shown that the advantages of teletext is that its pages are available before and after a programme, so they can act as notes for the teacher beforehand and reminders for their students afterwards. Moreover, young children can use the pages at their own speed and the pages are there whenever they want them. Experiments are being undertaken in five schools in California with station KCET, Los Angeles, based on a 55-page teletext magazine called 'Think Shop', comprising question and answer material and back-up information for teachers. It is transmitted twenty-four hours a day. Classrooms of the future will also have a large-screen electronic blackboard for group viewing, and hard copy will be obtained from an associated printer. There will also be a central library of prerecorded programmes and interactive video discs, which will be used for individual 'programmed' learning sessions.

If these developments become cost-effective and acceptable to teachers, the question will be who will pay? At a time of cuts in educational budgets in the States and the United Kingdom, together with inflation and rising costs, public funds are unlikely to be available. It is possible that the cost of production and distribution of television programmes and services and the price of computers will drop in the next few years. Industry and commerce will stimulate the development of educational systems for their own training programmes and may actually sponsor programmes; for example, in 1981 General Motors put up $1 million for a weekly series linking current events to basic studies in American high schools.

It is important to recognise that there is also potential income from the growth in the worldwide market for educational packages – especially for the United States and the United Kingdom whose programmes are made in the English language and are readily understood in most countries of the world. In countries where education has had to compete with general programmes for broadcasting time, the opening up of new outlets in satellites, cable and home video will present a variety of educational opportunities. On the other hand, a proliferation of outlets on the American scale could pose threats. It could encourage government intervention; it could lead to regulation of educational television in one form or another; at the other extreme it could lead to commercial exploitation and, perhaps, to a lowering of standards. In the United Kingdom, where educational broadcasting has been dominant for nearly sixty years, the possible threats and opportunities will raise important questions for the broadcasters, the educationalists and the educational councils.

RELIGION

In the United Kingdom, neither the BBC nor the ITV companies are legally required to broadcast religious material. However, both organisations do so. They also subscribe to a single Central Religious Advisory Committee (CRAC), which represents Christian and other faiths. Other specialist committees reflect regional and denominational interests. In the year 1980–1 nearly 300 hours of religious television programmes were transmitted, divided almost equally between the BBC and ITV and representing some two per cent of total transmission time. Programmes fall into two categories: programmes of worship, broadcast mainly on Sundays; and general programmes about religion, broadcast at other times, though seldom in peak hours. All costs are paid by the broadcasters, and no advertisements are allowed during the programmes.

As far back as 1925 the BBC agreed that no religious services would be broadcast during regular church hours, so as not to affect church attendance. (For the same reason no children's programmes were broadcast on Sunday afternoons during Sunday School time.) This protective attitude persists, despite considerable deregulation of transmission controls by the government in 1959 and 1972. The two organisations, in agreement with CRAC, still guarantee at least thirty-five minutes of religious broadcasting on Sundays from 6.40 to 7.15 pm on BBC-1 and ITV. By the same token BBC-2 normally broadcasts nothing that would compete.

Over the years the broadcasting organisations have pressed for a more enlightened policy towards religious broadcasts. Programmes now range wider than the work of the main churches and reach out to people outside the established faiths. Some programmes deal with non-Christian views of the world. But despite this more liberal approach, religious broadcasting still has to compete against general programmes within the various programme companies for its share of air-time and money. There are other independent producers, working in film companies and video production houses, who are free of this constraint and the policies of CRAC. A number of organisations are now active in the video field: The Christian Television Association; Christian Video Outreach; Campus Crusade for Christ; Audio Visual Ministries.

The Catholic Church has its own radio and television centre at Hatch End, just north of London. Its emphasis is on television training rather than on the production of programmes. CTVC, the Centre for Television and Radio Communications at Bushey, nearby, is similarly concerned with training, but also produces programmes on religious matters, social concern, health and education. The Centre also has a library of 300 programmes for hire. The ways in which teletext and viewdata services can be exploited for Christian information and education is the subject of a research project at Durham University, for which the Society for Promoting Christian Knowledge has contributed £45,000. Question and answer programmes on religious arguments are envisaged. One completed programme deals with the arguments for and against the existence of a personal God.

While independent companies can do what they wish with their own money on film and video cassettes, they cannot buy time on the established television networks. Regulation may initially have restricted the development of religious broadcasting, but it has kept a balance and prevented the worst excesses of present-day American religious broadcasting.

It is estimated that at least fourteen million Americans watch at least one religious programme a week. Over thirty television stations broadcast large amounts of religious programming and four religious networks are offering programmes via satellite to television stations and cable systems throughout the country. This so-called 'Electronic Church' leans towards evangelism. An estimated fifty million people in America are associated with evangelical churches or fundamentalist movements. The size of the 'Electronic Church' has made television preaching into big business. Contributions of the equivalent of £150 million a year are made to eight 'Television Ministries' alone.

One of the most important and lucrative is CBN, the Christian Broadcasting Network. CBN broadcasts by satellite for twenty-four hours every day. It claims to reach twelve million homes on over 2500 cable systems. Initially an all-religious network, it is now moving towards family programming and already one-third has nothing specifically to do with religion. Family programmes do have a 'morally-positive' aspect but are not intended to be religious or 'preachy'. With the exception of the 700 Club, which is hosted by the CBN president, Pat Robertson, all CBN programmes are now commercially supported. CBN claims to exercise moral judgements on choosing sponsors, and exclude all forms of liquor advertising, for example.

Unlike CBN, the PTL Television Network is supported totally by contributions from viewers. It distributes its programmes by satellite twenty-four hours a day to 320 cable systems, with a potential viewing audience of 2.7 million people. PTL itself only produces one programme, 'The PTL Club'; all other programmes are provided by other 'Ministries' on video tape.

Trinity Broadcast Network claims to be the broadest-based. They have programmes by Catholics, Methodists, Baptists, Pentecostalists, and other religious groups. Their programmes range almost as wide as the commercial networks but they all have a religious flavour. One of their game shows consists of schools competing against one another in answering questions about the Bible – a noteworthy exercise since religion is not taught in state schools. The Trinity Network currently reaches 150 cable systems with one and a half million subscribers. They also have four broadcast stations and are investing heavily in earth stations to extend their satellite network; their current investment is nearly £1 million. Trinity supports itself entirely through public contributions.

The Episcopal Television Network is the most recent venture. It is on the air for two hours on Sunday mornings and is distributed by satellite to 100 cable systems, comprising nearly one million subscribers. The two hours consist of a children's programme, Bible study, a news report and a church service. The Network claims that they are the first of the main Churches to reclaim their place in religious television and to cease depending on public access for broadcasting time. They do not accept advertising and are supported by gifts and donations.

Another recently formed network is the Catholic 'Eternal Word Television Network'. The Network is on the air for four hours every evening and is distributed by satellite to fifty different cable systems throughout America. They call themselves a 'holistic' network and combine entertainment and family teaching programmes with a high moral attitude. A typical evening begins with children's programmes, followed by folk or religious music, and discussions dealing with different forms of Catholic theology and philosophy. Financial support comes from both Catholics and non-Catholics; the Network has no commercials. Occasionally they solicit funds over the air, but they claim that there is no 'hard sell'. On-air soliciting can raise problems if a network is relayed on public access channels which do not allow such practices.

The National Jewish Television Network based in New York is another example of how cable distribution can be used by religious groups. By making use of satellites, the Network can link up forty-two different cable systems, representing well over one million subscribers. They produce eight half-hour shows a week, ranging from entertainment to programmes about the Yiddish language and culture. They are funded by private grants, but they plan to include regular advertising.

In terms of scale, the Mormons are planning what is said to be the largest satellite television network in the world. Five hundred satellite earth stations will link Mormon communities throughout the United States. They will be able to be part of the church functions in Salt Lake City and benefit from the proposed educational programmes. The network should be in operation in 1982.

The growth of televised religion in America is of increasing concern to traditional churchmen. There is a belief that the 'breadth' of the Gospel cannot be reflected in the so-called 'Electronic Church'. It becomes constricted by the medium, changing to a consumer religion of instant gratification. It does not help people think, but draws upon their feelings of disappointment and alienation.

Equal concern was expressed early on by churchmen in Britain about the effects of commercial television. It can too easily reflect and stimulate, so it was said, a glossy materialistic view of life, as it implies that consumption is happiness and that the accumulation of worldly wealth inevitably leads to a full and satisfying life. In the light of the American experience, what then will be their reaction to the development of cable and satellites in Britain?

Religious television on the American pattern does not appear to have problems over funding. However, how many established British churches will extend the idea of the collection plate to funding television programmes is not easy to assess. With direct broadcasting by satellite both the resources and the funding may come from other countries. Satellites may also be used to link together religious minorities in a number of European countries, like the National Jewish Network in America. Fervent religious minorities are likely to fund such operations. This will inevitably raise questions about the future role of the established churches in British television. Will they be content to be guaranteed television coverage as at present, or will some of them follow the lead of the American Episcopalians and buy television time if satellite and cable resources are available. If some were to follow the American pattern, this would raise serious policy questions for the broadcasting organisations and CRAC, and possibly for the government.

INDUSTRY AND COMMERCE

For many years industry and commerce have exploited television in one form or another. In doing so they have also provided the stimulus for the development of a whole range of new equipment of a non-broadcast standard. Cameras and video recorders designed for television broadcasting were too expensive and too sophisticated for industrial and commercial use. Japanese technology provided the answer in the form of simple, cheap electronic cameras and recorders. The original U-matic video cassette machine, for example, was developed primarily for industrial use. The growth of a non-broadcast industry has had repercussions on established television broadcasting, simplifying equipment and bringing down prices. Many items are now interchangeable.

The new equipment was used initially by industry and commerce for training programmes. Research had shown that employees liked video presentations because they were familiar with television, while managements liked television because of its flexibility. In Britain Barclays Bank were one of the innovators, and still use their comprehensive television installation for training of all kinds, including role playing. Another British bank, National Westminster, recently spent £2 million on a video communications network for its 3200 branches, with the object of abolishing paper as the medium of information. British Home Stores has installed video systems in its 120 shops. Marks and Spencer have transferred all their existing film and tape/slide programmes to video. Like British Home Stores they rent their equipment. One rental company has put together a package in association with a video club for hairdressers: a TV set and video recorder can be hired together with a choice of one-hour cassettes on new hairdressing techniques and the latest styles.

Some British companies have developed television as a means of improving internal communications. Shell UK, for example, has thirty TV monitors placed in its London headquarters for a continuing teletext-type programme of company news, called Shellfax. The BBC has a similar installation at its Television Centre. Other companies mount major television productions. When the Queen inaugurated the new oil terminal at Sutton Voe in the Shetlands, the British Petroleum Company hired television cameras and editing equipment. A programme about the opening was edited during the afternoon and shown to the 5000 workers the same evening.

There are similar developments in America. General Motors has installed 11,000 MCA Discovision players in its showrooms and work areas for customer demonstration, for employee training and other uses. The machines are the industrial version of the Philips' Laservision and embody all the same interactive elements. Ford Motors have bought 4000 Sony industrial players for installation in showrooms for sales and service training, as well as for management communications and point-

NEW SOURCES OF PROGRAMMES AND SERVICES

of-sale displays. IBM, who had a half share in Discovision, launched an after-sales training programme in computers based on the interactive video disc.

Another growth business in America is 'teleconferencing', which makes use of television to link up businessmen, thus saving costs and travelling time. This can be a small meeting within a company with minimum resources, or can be worldwide and make use of television crews and professional expertise. In America one of the largest teleconferencing companies owns 268 earth stations in 200 cities throughout the country. Hotel chains have recognised the business potential of 'teleconferencing'. Holiday Inn Inc. have hotels with permanent satellite dishes making up a network. In 1981 their work ranged from an all-day teleconference for Allied Van Lines for its 1500 agents and their employees in twenty-eight different cities, to the use of the network by the National Association of Realtors to teach new techniques of mortgage financing to 5000 realtors in ninety-five locations. An alternative to permanent earth stations is 'transportables'. For a US Army teleconference involving twenty-eight cities, fourteen transportables were used.

British teleconferencing: Glasgow and Euston in London linked by British Telecom's 'Confravision' system.

In the United Kingdom British Telecom is planning a series of trials involving fifty companies. It is funded by a £58,000 grant from Aregon International, the company which developed Cyclops in association with the Open University. Cyclops allows graphics and other information, written by a 'light pen', to be transmitted over the public telephone network for display on an ordinary television. The aim of the British Telecom trials is to find ways of using telephone circuits for low-cost teleconferencing. Cable and satellites are also being used to link electronic word-processors within companies or between companies in the same city or anywhere in the world, as a form of electronic mail.

In Europe this has the backing of the postal authorities and the system has been named 'teletex'. This is not to be confused with 'teletext', the one-way broadcast news and information service, nor with 'telex', the traditional cabling system, which it rivals. A two-page letter of 2000 characters can be sent in fifteen seconds by teletex, compared with five minutes by telex. Teletex is also being developed to transmit images. This will enable documents which contain graphics as well as text to be sent. A teletex system is already operating in West Germany, to be followed by one in Britain in 1982. There are city systems in America, and plans for a nationwide service. It is forecast that teletex will start to grow rapidly in the next few years and reach four million items a day in Western Europe and two million items a day in the United States by 1987, when it is expected to have eclipsed telex as a means of transmission.

Prestel, the British viewdata system, has opened its own Electronic Mailbox whereby subscribers can send messages directly and instantaneously to one another. A user who wants to send an electronic message calls up a 'mailbox frame' on his television screen. If he has a computer keyboard, he can write fifty words to the frame. A user with a cheaper, numbers-only keyboard can call up one of a selection of pre-designed frames and fill in the blanks with the appropriate numerals. The sender enters the account number of the receiver, usually the telephone number, and the latter's name appears on the screen. Finally, he presses a button to transmit the message through the telephone system. The message will be stored along with others in the receiver's television set until he calls them up. It is estimated that between 100,000 and 250,000 messages will be sent during 1983.

Nearly ten years ago the President of the RCA Corporation predicted that businessmen of the future would communicate instead of going to work. With teletex, teletext, viewdata, home video, and the growing interest in satellite and cable, the vision is coming close to realisation. For £50 you can now buy a British basic home-computer keyboard that plugs into your television set and works with recorded computer programmes. The BBC, in partnership with Acorn, a computer company, has produced television programmes and handbooks for a home computer which costs from £225 to £335. (The more expensive models interconnect with the teletext and viewdata services and are capable of receiving computer programmes from a variety of sources.) There are also plans to initiate an experiment in tele-writing for home business purposes, using the Cyclops equipment. About fifty companies will be invited to partake in the experiment in 1983. Currently British Telecom is providing £85,000 towards the Open University's own experiment with Cyclops as a teaching aid.

Home computers coupled to the television set are opening the door on electronic shopping. It is the view in America that the era of widespread 'telecommunication shopping' is now approaching, and that the very concept of retailing as well as the nature of retail competition will change radically. It is believed that customers are increasingly reluctant to spend time inefficiently shopping for goods in already crowded stores and supermarkets, when instead they can order many of the same goods by catalogue or telephone. Electronic shopping is expected to appeal to professional and management households to begin with, since they are likely to have the technical sophistication and the purchasing power to operate the new ordering systems.

Electronic shopping will involve traditional retailers, brand-name manufacturers, broadcasters, computer manufacturers, telecommunication suppliers, banks and credit card companies. Customers will shop at home using video-display catalogues, similar to the Sears Roebuck video catalogue already in use in shops in America and in the home for mail orders. The customer may also select from daily shopping guides on cable or broadcast television. Two such programmes started on American cable in 1981, *Home Shopping Show* and *Home Shopping Channel*. Each operates in 30-minute segments which are updated during the day. Customers have the use of a freephone number. *Home Shopping Show* is distributed by MSN cable to 456 systems throughout the country and is available to almost five million homes.

Orders are assembled by computer, and retailers pack the goods in fully automated warehouses. Customers choose between collecting orders from a nearby distribution point, or paying extra for delivery to their door. Some customers will shop electronically for food stuffs, especially groceries, others for consumer items, adding to the volume of present-day mail-order and phone-order business. Electronic shopping will be the practical and logical extension to the television commercials, in that it puts the point of sale into the viewer's hands in his own home.

Home banking via interactive cable systems or via viewdata is being considered as a method by which banks can reduce the cost of doing retail trade. Research in America suggests that consumers would be willing to pay a few dollars a month for the convenience of paying their bills through their TV set. Security would be maintained by the use of a banker's identity card, with a microchip embedded in the plastic at the time of manufacture identifying the account number. Meanwhile, as a first step in the United Kingdom, tests are being conducted by Barclays Bank in operating an Electronic Fund Transfer system at retail outlets, making use of their existing Barclaycard.

New television services such as electronic shopping, banking and mailing will initially be in competition with existing programmes for 'time' on the TV set. In the long term, though, the introduction of the home terminal will provide a separate monitor display for the television services, possibly located by the telephone. Nonetheless, television services and programmes will have to share resources with business interests who want to make use of cable, satellites and video. Telephone networks and two-way cable installations will also be developed for the transmission of data associated with the home – security services, fire, burglary and medical alert alarm systems. In America there are a dozen or so systems provided by Tocom Cable, in addition to the two-way QUBE installations pioneered by Warner/Amex in Columbus, Ohio. The home terminal adaptors cost about £75, with a monthly rental charge of between £6 and £9. The installation can be adapted to control energy consumption and to take meter readings.

To meet future demands in America, the FCC has approved plans by nine companies to build and launch more than twenty communications satellites by 1985/6. This will provide something in the order of 900 transponders, three times the 1981 capacity. In Europe, telecommunication authorities are developing cable networks and communications satellites for business use. The market value is reckoned to be only ten per cent of the American market, but the annual European growth rate will will nevertheless be over sixty per cent, changing shape as it does to approximate to the American pattern.

British Telecom plans to launch a satellite business communications service in 1983. A number of private companies are studying the possibility of introducing one or more rival systems involving cable and satellites. One consortium has been granted a licence to set up a new commercial network, called Mercury, in competition with British Telecom. Another company is working on a technique to use the electricity mains to transmit computerised information. It would also be technically feasible to run optical fibres the lengths of existing electricity cables. The Mercury project will be based on a fibre-optic network laid alongside British Rail tracks, linking seven major business centres in Britain. The consortium is made up of Cable and Wireless (forty per cent), British Petroleum (forty per cent), and Barclays Merchant Bank (twenty per cent). It hopes to link its first customers by the middle of 1983. The system will carry phone conversations, data transmission and teleconferences. It will also be technically possible to distribute television programmes.

In most countries there will be problems of regulation and funding and in Britain a difference of view between the two political parties about the degree of public and private investment. Business demands for television resources are going to grow at a very fast rate. The television industry will have to ensure that the expansion is not to its disadvantage. History must not be allowed to repeat itself. In the 1960s television was the poor relation of the telecommunication industry in terms of priority in the use of European circuits and transatlantic satellites. With systems of direct charging, such as pay-tv, television should be rich enough to buy its fair share of the new communications spectrum.

6 THE CHALLENGE TO BROADCASTING

In America a hardback book is a best-seller if it sells between 100–500,000 copies. A paperback is a best-seller if sales reach one to ten million copies. The two newspapers with the largest circulations, the *New York Times* and the *Wall Street Journal*, each sell about two million copies a day. A successful film will draw six million patrons. *TV Guide*, a best-selling magazine, sells around twenty million copies a week. A television programme can attract twenty million viewers, but this may not be enough to keep it on the air.

Competition between the three commercial networks in America for ratings, and thereby for advertising revenue, has reached these phenomenal proportions to the detriment of the quality of programmes. There is a tendency to reduce entertainment programmes to the lowest common factor, so much so that, according to a Gallup poll, only thirteen per cent of viewers were 'very satisfied' with national network output; forty-seven per cent expressed relative satisfaction, whereas forty per cent said that they were little satisfied or not satisfied at all. The latter group was made up mainly of the better educated – college degrees and above – and those in the higher income bracket.

One forecast suggests that, by 1985, about fifteen per cent of the time spent using the television set will be devoted to the viewing of non-broadcast television programmes. Assuming that the amount of viewing time remains as at present – six and a half to seven hours a day – this could mean about one hour would be lost to the American networks. What the one hour would consist of is not easy to predict, but given the availability of the technology in the home, it will be the nature of the programming that will determine the choice.

Comparison has been drawn between ABC, CBS and NBC and the great titles of the print industry, *Look*, *Life* and the *Saturday Evening Post*, which proved to be the dinosaurs of the television age. It is argued that the networks will be undermined by the new distribution systems because they are not only cheaper to operate but offer the consumer a wider choice of programmes and services, which he pays for directly. With this in mind some observers forecast that audiences of the three commercial networks could be cut in half by 1990, or one or other could close down, as a result of changes in viewing habits, or could, like radio, change to generic scheduling.

Broadcast television in the United Kingdom is, on the other hand, actually on the increase. Channel Four is scheduled initially to broadcast for some sixty hours a week. Regular breakfast television, which starts on both BBC and ITV early in 1983, will account for a further thirty hours a week. The Welsh Channel Four also comes into being towards the end of 1982, with an additional twenty-two hours a week in Welsh.

Schedules will be more balanced than in America, in line with existing networks. As an example, the output of the BBC's two national networks in 1982 can be rounded up as follows (excluding the Open University): News 4%; Features 20%; Sport 17%; Education 10%; Children's Programmes 9%; Light Entertainment 7%; Drama 6%; Religion 2%; Music 2%; Feature Films and Series 18%; and other material 5%. The proportions for the individual ITV companies cannot be so easily expressed, but the overall pattern is similar to BBC output. Channel Four plans to have a slightly different balance: News and Features 25%; Drama 10%; Education 12%; Music and Arts 15%; Feature Films 25%; and the balance made up of sport, children's, youth and other material. Up to twenty per cent of their programmes will be commissioned from independent producers.

There is no evidence yet of viewers turning away from broadcast television because of dissatisfaction with the programmes, as in America. Viewers with teletext temporarily switch away for news or information. Homes with video-cassette players use the television set for playing back programmes. The limited pay-tv experiments have, however, shown that new or blockbuster feature films will attract a household from broadcast programmes and that the viewers are prepared to pay for the service. More inroads may be made into the broadcast audience when the two DBS services are introduced in 1986; by then, a number of urban cable systems should have been installed to bring these services into more homes instead of relying on DBS dish aerials for every household. If that happens, and it is declared government policy, then other programme services are likely to follow, in the pattern of American cable.

So development of television in Britain may begin to resemble television in the United States, as far as distribution technology is concerned. In the face of this the BBC, for example, has made sure that it has gained some foothold in all the new markets, in addition to maintaining its broadcasting obligations. It was the first to develop teletext services. It already has a commercial subsidiary, BBC Enterprises Ltd, which not only sells its programmes to other broadcasters, but now also seeks outlets in the home video and cable markets both in the United Kingdom and overseas. BBC Enterprises has made an agreement to supply forty per cent of the programmes for the new American RCTV Entertainment Channel. The video cassette of the Royal Wedding in 1981 sold 5000 copies in the first week alone, even though viewers could have recorded the ceremony off air. The BBC is also the programme supplier for Visionhire, one of the seven companies granted a licence for the two-year pay-cable experiment in Britain. Showcable, as the service is called, is making use of the BBC's skill as a programme scheduler and its experience in acquiring programmes on the world market. For the experiment the only BBC-originated programming will be its CEEFAX service. The BBC formulated plans early on for satellite television and was awarded the licence for the two DBS channels scheduled to start in 1986 – one a pay system based on new feature films, sport and cultural programmes; and the other, the best of television, *Window on the World*.

In America the three commercial networks have similarly diversified their operation to gain access to the new markets. All three formed commercial subsidiaries to sell programmes into the home video market: ABC in July 1979, CBS in December 1979, NBC in January 1980. Each is interested in overseas sales; CBS, for example, released their video cassette 'The Hostages: From Capture to Freedom' in the United Kingdom, and ABC are distributing a weekly news magazine for Americans overseas. All three, or their parent companies, have also set up satellite cable networks; CBS Cable in October 1981, ABC ARTS Network in April 1981, and NBC's parent company, RCA, starts its Entertainment Channel in 1982. In 1982 ABC, in association with Hearst Newspapers, will introduce a four-hour daily cable service for women, and they are planning two 24-hours-a-day satellite news channels, one of which will make use of ABC's network news resources. CBS is concluding teletext experiments and has submitted proposals to the FCC for an integrated teletext service.

The commercial networks in the United States are thus in a new competitive situation. Competition is no longer just network to network for the maximum share of the audience but between the networks and the new non-broadcast services. The situation is further complicated by the networks themselves competing for audiences in the new markets. It is not merely a ratings battle. There will be competition for both programme material and for programme funds. The situation will become critical by 1985, when the overall income of the non-broadcast systems will probably exceed the income of the networks. This will put the networks at a disadvantage in bidding for television rights for events not only in the States, but elsewhere in the world. With this in mind ABC, for example, have made a deal with ESPN to share the costs of covering the Olympic Games in 1984. Different versions will be broadcast and distributed by satellite for cable and closed-circuit television and also sold as video cassettes and discs. It is intended to be a co-ordinated and not a competitive operation.

In the United Kingdom there is as yet little competition for audiences between broadcasting and the non-broadcasting services, as they are still in their infancy. Competition between broadcasters, however, has already extended to the new markets. In home video, for example, the BBC and Thames both produced cassettes of the Royal Wedding. The BBC and the ITV companies are also competing for exposure on the new cultural cable channels in America. From 1986 onwards competition may become more serious as the new markets in the United Kingdom rival domestic broadcasting in some programme areas, especially feature films and sport. There is already concern among broadcasters that they may lose access to major sporting events in the face of higher bidders from the non-broadcast sector who may be funded by entrepreneurs. It is in the broadcasters' interest to come to some form of partnership arrangements, to safeguard their future access to such events.

This, however, is not the total equation. The potential competition from associated industries has to be added. Radio will remain an effective competitor to breakfast and daytime television, in terms of audiences and money. The record industry has a hold on musical talent, especially pop, and is already making its own television programmes for home video and for other forms of distribution. For the film industry, releasing films to broadcast television will be low on their order of distribution priorities; home video and pay-tv will come first. Theatrical performances will be bought up exclusively for home video and cable where specialist interests can be catered for more economically. Newspapers, magazines and books will be in competition in the production of specialist programmes and services, such as teletext.

Industry and commercial companies will be the providers of other services to the home – shopping guides and electronic mail, for example. They will also be competing for resources such as cable and satellite channels. Educational and religious organisations will no longer be subject to the constraints imposed by the broadcasters and will be free to produce and distribute their own programmes. As in its dealings with sporting organisations, the broadcasters should initiate partnerships or make some mutually beneficial associations with these industries and institutions, before they become inextricably and exclusively bound up with non-broadcasting organisations, or they themselves become non-broadcasters in their own right.

If broadcasting organisations, in the face of these threats, are to survive, they must increase their distribution outlets and at the same time enter into new production partnerships. From a position of such strength it will be possible for them to arrange for the orderly and commercially viable distribution of programmes, balancing the needs of the popular mass audience with those of specialist and minority-interest audiences. It is up to the broadcasters to take the initiative, and to seek the means of funding.

7 MEANS OF FUNDING

Traditionally television broadcasting throughout the world has been funded by a variety of means: licence fees, government grant, the sale of advertising and programme sponsorship, or combinations of these. Licence fees, although they protect the independence of the broadcaster, are determined by governments. Broadcasting organisations financed directly by governments also have their income levels determined by the state, but have little or no editorial independence. Broadcasting organisations supported by advertising, though tacitly independent, rely on high ratings and the buoyancy of the advertising market. When programmes are sponsored, the broadcaster gives up his editorial responsibility to the sponsor, in return for a high fee.

In Europe some broadcasting organisations are funded only by licence fees, which currently vary from £72 a year in Denmark, the highest, to Sweden £64, Norway £56 and the United Kingdom £46, the lowest. On television in these Scandinavian countries as on the BBC in Britain, there are no commercials. The ITV companies in Britain are funded totally by advertising and are allowed a maximum of six minutes in any hour. Other European broadcasting organisations are funded by a mix of licence fees and advertising. The amount of advertising allowed is more strictly limited. In France, Germany and the Netherlands only twenty minutes per day on each channel is allowed; the balance comes from licence fees.

Coproduction money may be attracted to help with the cost of major programmes. The idea of trading distribution rights for a cash contribution to production costs by other producers or distributors is becoming more difficult with less money available and demands for editorial involvement by the coproducers. Broadcasting organisations also sell their programmes to other broadcasters and non-broadcast outlets, as well as exploit associated books and merchandising.

Irrespective of the source of their income, the way in which the broadcasters spend their money to provide television and radio services, is broadly similar throughout Europe. The BBC, for example, provides two national television services, four national radio networks, regional television and local radio. It pays all production costs for its programmes as well as owning its own production facilities and transmitters. It pays rental to British Telecom for the use of circuits. In America the three national networks, ABC, CBS and NBC, are funded by advertising and sponsorship of programmes. The PBS network and independent television stations derive their incomes from a variety of sources, both private and public, though alternative means of funding are being considered – commercials or over-the-air subscription television – to offset the cuts in federal funding of public broadcasting. The pattern of spending on production is broadly the same as in Europe, although a far higher proportion of programming is bought in from independent production houses. The national networks and their affiliates own the transmitters which are linked by circuits leased from telecommunications companies.

Non-broadcast systems also earn their income from a variety of sources. The satellite programme companies in the United States, on the one hand, act as 'wholesalers' for cable and pay-tv stations and other outlets, distributing programme packages. Their income comes from selling advertising, sponsorship and the rental charges paid by the stations relaying their programmes, usually based on a few cents per subscriber. Initially the programme packages, mainly feature films, were bought in, but now more and more satellite programme companies are distributing programmes they have produced themselves. The important distinction is that the companies do not themselves own the satellites – they pay rental charges for the hire of the transponders.

Cable stations in America, on the other hand, act as the 'retailers' for viewers. They earn their money from local advertising and from a rental charge paid by their subscribers. They offer a choice of local broadcast stations, community and public access programmes, local language programmes and a selection of packages distributed by satellite. A cable station will own its own cable network, receiving equipment and a dish aerial for satellite programmes. The station may be involved in the production of local programmes, but usually they work in association with local institutions.

Pay-tv stations in America can be regarded as the 'up-market retailers', making an additional charge for their exclusive programmes. At present one of the attractions of pay-tv is the absence of advertisements, but as its penetration of television homes increases there is pressure from the advertising industry for access to these more specialised markets. RCTV's Entertainment Channel, for example, could become the first pay-tv service with advertisements. Subscribers are charged according to the package they choose – either on a pay-as-you-view basis or for the complete output of the channel. Pay-tv channels currently consist mainly of new feature films, sport and one-off 'spectaculars' produced by the pay-tv company. Pay-tv makes use of existing satellite and cable distribution systems.

There are also a growing number of over-the-air subscription television stations, which originate their own programmes, buy in packages from other distributors, or relay one or other of the satellite services. There is an FCC proposal to licence a number of low-powered transmitters with only about a ten-mile radius, which would operate on a subscription basis during prime time and provide free outlets for minorities and other groups not yet catered for at other times. At the beginning of 1982 the FCC had already received 6000 applications. As far as can be seen DBS systems will also operate on a subscription basis. The initial audiences will not be large enough in themselves to attract advertising.

The distinction between pay systems and broadcast and cable television is that pay systems are funded *directly* by the consumer, who pays for what he wants to view. The same is true of the home video market. A home video consumer has first, of course, to buy his videocassette or disc player in order to 'distribute' programmes to himself. He also has to buy blank tape to record programmes off-air or for home movies, or has to rent or buy individual programmes at a cost which would give him several months' television from a pay-tv service.

There is also an important distinction between the broadcasting organisations and the new non-broadcasting companies. The broadcasters own and control their transmitters; the non-broadcasters do not. They lease time on other people's systems, frequently on a franchising basis. Because of this the American networks, while retaining their broadcast systems, have also been able to move into the new non-broadcast markets by leasing facilities at little or no capital cost. The original cost of cable installations was paid by the cable companies. Satellites are owned by large telecommunications corporations. They were launched as an investment for business communications and are now used increasingly for television. Future expansion of cable networks is likely to be funded by successful pay-tv services and by business interests.

In Europe the emerging pattern is not dissimilar. Broadcasting organisations are able to enter the non-broadcast markets by leasing resources with very little capital cost. For example, the satellite planned for DBS over the United Kingdom is being funded by private money from a consortium including British Aerospace, Thorn EMI and Rothschilds. The BBC will lease two transponders. In Germany and France, on the other hand, government money will fund the building and launch of their planned satellite. The BBC expects to earn enough from the one proposed subscription channel to pay for the programme and rental costs of both channels.

Similarly extension to the existing cable installations in Britain is being financed by private investment, mainly because the investors want to extend the business communication network throughout the country. As in America, cable services will be shared between television and business users. It is also possible that British pay-tv companies will in the coming years be able to help fund new cable installations.

In Britain home video, like pay-tv, is directly funded by the consumer. The companies involved are investing in other forms of distribution. For example, Thorn-EMI, who are established as film and television producers and distributors as well as being manufacturers of home video players, are investing in the British satellite and are one of the foremost companies developing programmes specifically for the home video market. In America CBS as a television company and Twentieth Century-Fox as a film company have formed a joint venture to develop, manufacture, market and distribute cable and home-video programming.

In Europe broadcasters will be able to develop new services on a directly funded basis that will meet the requirements of minority and specialist-interest groups more satisfactorily. Expensive productions will be funded by going into partnerships with sports organisations and artistic institutions and finance companies, on the pattern established for Covent Garden. These new elements will free broadcasting organisations to some degree from their present financial dependence on advertisers, sponsors and government.

8 TECHNICAL QUESTIONS

All television has a technical base which places constraints on the production and distribution of programmes whether by broadcast transmission, satellites, cable, video cassettes or video discs. Twenty-five years ago, a producer could not cut or mix from one programme source to another; most programmes were broadcast 'live'; the only means of recording was on film.

Magnetic tape has replaced film. CBS in America, for example, now operates totally without film resources of any kind in its studios, on outside locations and for news. This has enabled the company to streamline its operations and to develop sophisticated, computerised post-production facilities. UHF transmission has replaced VHF, opening up more broadcast channels as well as introducing colour television, which requires more 'space' than monochrome. Satellites now link television across continents – twenty-five years ago urgent news film could be sent slow speed across the Atlantic by telephone circuits – a poorly graded and costly operation. Some original cable installations in Britain can still only carry four channels, whereas coaxial networks can distribute up to seventy-two and can cope with two-way television. The next generation of fibre-optic networks will be capable of carrying hundreds of programmes and services. The last decade has also seen the sophistication of video recording for professional use, or for use in the home – the video cassette and the interactive video disc.

It is important that these new facilities are used efficiently and economically. In America full television cable channels are being used for simple information services, when far more comprehensive information can be stored on a few teletext pages and piggy-backed on television signals. Using cable for two-way information can also be wasteful – the telephone network is cheaper and each subscriber can be linked to every other subscriber for speaking and for carrying simple visual information. It is, of course, through the telephone network that viewdata systems link the television set to central computer stores. It would be uneconomic to operate viewdata on a cable network.

In some circumstances an interactive video disc may, in fact, be cheaper and faster for retrieving information than using viewdata systems. As publishers begin to transfer their reference books on to video discs, it may be more economic to buy a disc than the actual book. Moreover, it would be considerably cheaper than making, say, twenty calls a day to retrieve the same information from British Telecom's

Prestel service, where you have to pay for the call and for each page.

The economics of DBS raise other questions. The cost of a dish aerial and tuner, for example, will be in the order of £200, for which you may be able to receive five domestic channels. It has been estimated that it would cost only £100 per household to double the number of cable subscribers in the United Kingdom, or £250 per household to install a full cable service into new areas. New installations would be coaxial or fibre-optic cable with the capacity of hundreds of channels. Moreover, the cable company would relay not only the four national networks and five satellite channels, but also a selection of foreign programmes and other services available with cable television.

By far the biggest technical problem is the lack of common world standards in television production and distribution. The audio record industry put its house in order in the 1950s when the industry adopted the standard 45-rpm single play and the 33-rpm LP. Records can be played on any player anywhere in the world. In television there is international agreement over the allocation of frequencies for both broadcast and satellite transmission, and on the area and placing of the national satellite footprints. There is, however, no agreement for teletext and viewdata specifications; digital recording; video-cassette and video-disc formats; or for television systems. The European Broadcasting Union (EBU) is active in bringing about agreements in Europe, on the specification for teletext and viewdata, for example, and have agreed a common standard for digital recording with the American Society of Motion Picture and Television Engineers (SMPTE). The three video-cassette formats (VHS, Beta and Philips) and the three video-disc formats (VHD, Selectavision and Laservision) will probably be settled in the long term by market forces.

If a world television standard were agreed, the problem of writing other technical specifications would be simplified. But any change in standards would require new production and distribution equipment for the whole industry and a new television set or converter for every television household in the world. Ideally, therefore, any new system needs to be compatible with existing systems so as to cover the period of overlap. The changeover could not be done overnight. It would probably require up to twenty years.

There is, however, a move towards standardisation in Europe. While each country is planning its own domestic satellite services, thirteen members of the EBU are participating in five one-week experiments during 1982, using the existing OTS satellite for a pan-European Television Service. If these pilots are successful in both programme and technical terms, it is hoped to start a regular European DBS service when the L-Sat satellite is launched in 1985. One of its five channels has been offered to the EBU for just such a service. The significant factor is that the IBA, which is the British partner – the BBC is concentrating on providing the two national DBS services – has proposed that the PAL and SECAM systems should be abandoned for satellite services and be replaced by the IBA's concept of a 'component' format which would make television compatible throughout the continent, possibly on a higher definition standard.

While this might unite Europe, if accepted, it would leave out America and Japan, the two key NTSC countries. CBS in America and NHK in Japan, together with some members of the film industry, are maintaining that, if there is to be change, it should be fundamental and lead to one world standard for television and for the film industry. CBS and NHK are jointly sponsoring a high-definition television system (HDTV) with stereo sound, based on 1125 lines and digital transmission. This would match the picture quality of 35-mm film. They also propose to change the aspect ratio of the screen from 4:3 to 5:3 to improve picture composition.

Even though the signal would be compressed 8:1 during distribution, HDTV would still occupy the 'space' of at least two colour television channels. CBS have submitted a plan to the FCC for an experimental HDTV service distributed by satellite. This raises fewer technical problems in America where national satellite channels are adjacent; in Europe national channels are alternated. Taking up twice the 'space' will mean fewer services, but the flexibility of digital systems and the enhanced 1125-line picture quality far outweigh that. Cable networks in any case will have hundreds of channels available. CBS believe a start should be made at the production end and that major television programmes and feature films should be produced and recorded on HDTV. No attempt should be made to make HDTV compatible with existing systems. Two versions would be made, with the 5:3 picture masked electronically to fit the 4:3 screen, and the signal down-converted from 1125 to 625 or 525 lines.

The film industry could be first to adopt a plan on these lines, distributing their recordings by satellite to video theatres, and leasing them to pay-cable networks. Producers of other types of material would follow. In terms of the home viewer, if he wanted the HDTV version he would invest in an additional receiver and big screen in much the same way as he does with hi-fi audio equipment.

Television broadcasting could be faced with the decision of whether to continue as now with PAL, SECAM and NTSC, or to accept a broadcasting solution based on enhancement or a merging of existing systems of 525/625 lines with the picture staying at a 4:3 aspect ratio. The only other alternative would be to follow the film industry and possibly cable and satellites, if they in the interim have adopted a high-definition system.

9 LAWS AND AGREEMENTS

Whilst there are conventions between all countries about the use of frequencies for television and radio signals, without which there would be chaos in the skies, the international agreements about creative rights – called by the lawyers 'intellectual property rights' – are more diverse. Most countries are signatories of the Rome Convention, the Berne Copyright Convention and the Universal Copyright Convention. These cover all aspects of broadcasting, but were enacted before the advent of the new technologies, which have introduced further international legal complications.

For example, countries do not agree on a legal definition of a 'videogram' – the collective term used to cover both video-cassette and video-disc programmes. Nor have they agreed under which current legislation a videogram should fall – that referring to television programmes or to films. The new systems are creating other legal problems affecting the distribution market. When a television programme is made in the United Kingdom, for example, an artist signs a contract for domestic showings; an additional fee, or residual payment, covers sales to other television stations abroad, with a token residual payment for non-theatric sales. A programme would be licensed for specific countries, when sold to another distributor, even for one country in Europe.

With the new systems such precise demarcation is no longer possible. New types of contracts are required to include videogram and pay-tv rights. No creative artists will now accept that non-theatric rights, if those include pay-tv, DBS and cable systems, should be at a token fee. The Actors Guild of America even went on strike to fight for higher payments to recover some of the money which the distributors were making from sales of programmes to the non-broadcast systems. DBS will cause problems of its own, in that the footprints from a satellite cannot be restricted to precise national boundaries, particularly when there are groups of small countries very close to one another, such as in Benelux. Any satellite service covering, say, Luxembourg will spill over into France, Germany, Belgium, the Netherlands, and parts of Denmark and the United Kingdom.

Each country currently has its own set of laws covering advertising which is allowed on its domestic channels. With the potential of pan-European DBS services, some of which will of financial necessity carry advertising, new conventions will have to be agreed. For example, in the Netherlands all advertisements for sweets have to carry a small toothbrush in the corner of the screen to warn both parents and children that sweets are bad for children's teeth. A decision would have to be taken whether all advertisements should carry such a motif, or whether the Dutch should adapt to different rules. If pan-European advertisements have one set of rules, there would inevitably be repercussions on national codes of practice. A case involving overlap of transmission in Europe has been heard before the European Court of Justice. It concerned a Belgian cable company which relayed a film from German television, although another company owned the rights for transmission of the film in Belgium. It was an important test case. The Court ruled 'that the unauthorised wire diffusion of broadcasts originating from another state of the EEC is not permitted'.

In America such problems are not so apparent. Apart from spillover to Canada and Mexico, the very size of the United States makes local frontier 'borrowing' of programmes irrelevant. Nonetheless, an important case has been going through the courts since 1976. Walt Disney Productions and Universal City Studios challenged the right of Sony to sell their video-cassette recorders, on the grounds that they were providing the means for the general public to steal the rights of the programme makers, and that Sony had made a machine that meant millions of Americans were infringing their copyright in programmes. The original decision of the court exempted home taping from copyright registration. An appeal has now reversed that decision, but further proceedings are in hand.

The fact that anyone can now record a programme or a film directly from his television set, without the original creators of that programme receiving any additional payment, is causing concern in most countries. Just as an author receives a royalty on each book sold, so film-makers and artists who have created a film feel that they should have some part of the money earned from the continuing sales of their films. In Germany it was decided that there should be a levy on the sale of video equipment, and the money collected distributed to the various artistic associations. In Austria there is a levy on the sale of blank tape. In Britain the various creative organisations have been pressing for a tape levy but the government is not in favour at the present time.

The overriding problem facing the industry, however, is that of piracy. It is a problem which the record industry has been fighting for many years. Not only do pirates make cheap recordings; they even copy the packaging of cassettes and discs and sell them throughout the world as the genuine article. The artists and the record company themselves receive no money whatsoever. The International Federation of Phonogram and Videogram Producers (IFPP) and the Record Industry Association of America (RIAA) estimate that millions of dollars are lost annually to their members from pirated recordings.

It is already known that the home video industry is suffering from the same problem. Copies of feature films newly released are available for purchase in London, although those films have not been officially released by the distributors. Video cassettes are usually of very poor quality, but purchasers do not seem to mind. *Star Wars*, for example, was on sale many months before the film was released officially.

As it is so easy to obtain material to sell on the 'black market' – all a pirate needs is a video-cassette recorder and a television set – there has been a growing trade to areas such as the Middle East. One cassette will be made in Britain and taken back to the Middle East, where other copies are easily made and then sold. The programme makers receive no payment at all. Some will say that part of the reason for this is that the material seen on television has not yet been officially released by the broadcasters, but they in turn cannot do so until agreement is reached with the many talent unions involved. At present it means that everyone is losing out and only the pirates are making money.

It is an international problem. Now that video-cassette recorders are readily available throughout the world there is a huge demand for programmes, particularly in those countries where national television and the cinema is either censored or strictly controlled. Initially the demand was mainly for 'adult' entertainment, but now there is simply a market for programmes of all kinds. Whilst attempts are being made in America and Europe to combat piracy there seems to be little success so far. In the United Kingdom the Video Copyright Protection Society, set up by the BBC and the ITV companies and representatives of the film industry, is trying hard to ensure that stronger penalties are handed out to those who are caught, and that safeguards should be enforced to prevent valuable film prints from 'straying' into the hands of the pirate copiers.

Another form of piracy is also growing in Europe. After the pirate radio stations of the 1960s there are now a growing number of pirate cable television stations. Once the cable installations exist it is fairly easy for a technician to plug into the system and distribute other programmes. In the Netherlands, where some fifty-five per cent of the country is cabled, there have been several pirate stations, most notably in Amsterdam where a pirate service feeds programmes into the cable network, once they have finished showing programmes for the evening. The pirates show mainly feature films and 'adult' movies, interspersed with local advertising.

Broadcast television in most countries is run on the basis of generally accepted standards of taste and decency. With the advent of the new systems there will be many opportunities for the unscrupulous to take advantage of the diversity of outlets to show programmes which could not be shown on broadcast television. Not only that, but those who have invested their money and talent in making programmes may find that they are no longer in possession of their legal rights, that others are exploiting their programmes without permission or payment. The coming of DBS means that international agreements need to be drawn up, so that acceptable regulations cover the transmission of programmes direct into people's homes. Most important of all, it is now that governments should look at the legal problems which the development of the new systems will bring, so that a workable and satisfactory legal framework can be created for the future.

10 GOVERNMENT REGULATION

Whereas the law and the courts constrain television production, governments regulate the whole industry. In 1923, a year after the start of British broadcasting, a government committee was saying that 'the control of such a potential power over public opinion and the life of the nation ought to remain with the State'. Within four years the government had delegated its responsibilities to the BBC, which it created as an independent public corporation, under Royal Charter.

Paradoxically, in the United States at the same time it was the broadcasting industry itself that appealed to the Federal Government to regulate the use of radio frequencies and to end the air-waves chaos. In 1927, the same year as the BBC's first Charter, the American government passed the Radio Act, followed by the Communications Act of 1934 which created the FCC, the Federal Communications Commission, to license broadcasting stations and to police the communications industry in terms of programme and technical standards. The way in which the FCC, acting for the Federal Government, is dealing with non-broadcast television and its attitude to the networks is relevant to governments and television organisations in the United Kingdom and in Europe generally. They, too, will be faced with decisions that will affect established relationships. Traditionally, the FCC and the courts had always taken a protective attitude towards broadcasting at the expense of the newly developing non-broadcast systems. The situation has now almost been reversed.

In terms of cable, for example, a cable operator may pick up and relay as many signals as he wishes from anywhere in the United States, even if the same programmes are being carried by local on-air broadcasters. He need not get the consent of, nor make any payment to, the programme company whose signal he picks up, though he may have to pay a small copyright fee to the owner of the programme. Cable operators are able to set their own subscription rates, and they can compete with telephone companies for provision of data and other services for the home. The FCC has also swept away practically all restrictions on pay-tv and over-the-air subscription television. There are now no limitations on the number of pay stations per community; on the amount or content of sport or feature films; on the nature of the scheduling; on pay-tv carrying commercials, if they wish; and one company can now provide both the equipment and the programming for a cable system.

The FCC still keeps a tight hold on the networks. It does this through their affiliates in that it allocates their broadcast frequencies on behalf of the Federal Government. The FCC, in fact, licenses all broadcast television stations, including the few owned directly by the networks as well as non-commercial and educational stations. The authority of the FCC is required by satellite companies to construct and launch satellites and for their positioning in space, as well as their frequency allocation. The FCC will also authorise the setting up of individual DBS companies, again because frequency allocation will be involved. Likewise, CBS will need the authority of the FCC for their proposed HDTV experiment and for stereo transmissions.

On the other hand, because cable does not call for frequency allocation in the same way as broadcast television, the FCC is not so involved. Permission to lay a cable network is a matter for local authorities, who award the cable franchises. It is then left to the cable operator to provide the services, but there is usually a commitment to the local authority to allocate, say, four channels for public-access television and relays of council meetings. The granting of cable franchises is highly political. In Minneapolis St Paul the choice proved so difficult that the local authority awarded the franchise to itself... and later changed its mind.

The awarding of pay-tv franchises is a responsibility of local authorities for the same reasons. So is STV, because the licence for the television station will have been granted by the FCC. The FCC invented a 'complement of four' rule which restricted STV initially to one per city. That is under review and is likely to be withdrawn so as not to limit competition in the new technologies.

The FCC has no jurisdiction over home data services, whether distributed by cable or telephone circuits, since no broadcast frequencies have to be allocated. The same is true of viewdata and the FCC is wary of being caught up in the conflict between federal and state telephone interests. Their unwillingness to arbitrate between the various teletext standards is a fair indication of the FCC's present attitude of letting market forces decide. This attitude, coupled with the steady deregulation of the non-broadcast systems, is evidence of their no longer being protective of the networks' interests. In fact, they even regard home video, for which they have no responsibility whatsoever, as being a healthy growth of alternative programming. The FCC cite radio as an example of the success of their policy. Radio is totally deregulated because there are enough stations to meet the widest needs of the market. Deregulation of television on the same scale will be possible when the market forces provide enough services to meet consumer demands.

The FCC claim to be ahead of Congress in wanting to do away with the Fairness Doctrine, which is concerned mainly with political balance. They believe there are already enough television outlets for 'balance' to occur naturally without their having to prescribe rules. Their only stipulation is under the Financial Interest Rule – there must be no cross-ownership between television stations and cable networks in the same town or between a local television station and a local newspaper. They can police the situation because they allocate the broadcast frequencies.

Despite the deregulation of non-broadcast systems, and moves by the FCC to reduce its responsibility for programme balance, there has been no relaxation in the Financial Interest Rule as far as the networks are concerned. This is important to the networks because it puts constraints on the amount of diversification they are permitted. When CBS, for example, produce a programme for their network, it can only be syndicated to other television stations through a third party for a flat rate. An associate company like CBS Sports within the CBS Group could not be the third party. If a coproducer is involved, the supplementary rights are his. There are no restrictions on a network selling or leasing the programmes it produces to non-broadcast outlets – cable, pay-tv and home video. On the other hand, if CBS buys in a programme instead of producing itself, it cannot be syndicated by the network. Associate companies, CBS Cable, for example, can buy other distribution rights, if separate approaches are made at different times. These provisions under the Financial Interest Rule apply to dealings in the United States and overseas. On each count, therefore, the networks are loaded against the new markets. It is a situation that disturbs the National Association of Broadcasters, as does the deregulation of the non-broadcast systems. The general belief is that the FCC will have to free broadcasting of its antiquated restrictions if broadcasting is to keep pace with the new markets.

In Britain there is no equivalent of the FCC. The Home Office deals with broad issues of policy including the allocation of frequencies, but in practice the government devolves responsibility for running radio and television broadcasting to the BBC and the IBA, once it has set out the ground rules as prescribed in the Licence and Charter for the BBC and the Parliamentary Acts for the IBA. This makes the BBC and the IBA independent in terms of domestic policy and day-to-day control.

Against this background, and in the light of American experience, how will the government regulate or not regulate the new television systems in the United Kingdom? The degree of government involvement depends on the political philosophy of whichever party is in power. It is the Reagan administration that has accelerated the deregulatory processes in America in the wake of the less active Carter administration. And it is the present Conservative government in Britain that has placed the new Channel Four under the IBA, reversing proposals of the previous Labour government to establish a separate Authority. On the other hand, the present government has created a separate statutory Authority, on a legal par with the BBC and the IBA, to run the new Welsh channel, even though it will be funded by the IBA, who will contribute £20 million in the first year, and the BBC, who will supply ten hours a week of Welsh programmes free, paid for out of the BBC licence revenue.

It is also the present Conservative government that has changed the BBC's Charter so as to allow the Corporation to take full advantage of new technology. The use of satellites is there by implication; subscription television is mentioned, and there is a reference to 'signals over wires' or 'other paths provided by a material substance', which could include fibre optics for the future. The BBC may also provide 'other services whether or not broadcasting services'. In other words, the BBC could become narrowcasters as well as broadcasters, but broadcasting remains the basic object and subscription television would be developed as part of it. By definition, broadcasting is intended from the outset to be received by everyone who possesses the necessary equipment; narrowcasting is intended for reception only by restricted groups. Narrowcasting will be what the Home Office calls services 'on the fringe of the BBC's traditional broadcasting services', and it may take many forms. The new Charter came into being in 1981 and will run to the end of 1996.

Following these changes, the government awarded the BBC the licence to operate the first two DBS television channels in the United Kingdom. These are planned to start in 1986. The satellite will also have six sound channels, which could provide not only superior 'digital' radio services but stereo television as well. Satellite channels will be awarded to the IBA at a later date.

The government is determined to bring about new legislation to encourage the fullest possible development of cable technology during its present term of office, and has appointed a three-man committee, under the chairmanship of Lord Hunt, to report on the implications of introducing cable television in Britain. The installation of a national cable network with the capacity of possibly thirty channels, potentially developing like American cable television, raises immediate questions about ownership and provision of programmes and services, as well as questions of funding and standards. It is precisely these questions which are under review.

The present cable companies collectively have just over two and a half million subscribers. Advertising is not allowed; their income is derived from rental charges. They may only relay existing broadcasting services, though there have been two experiments in extending the range of programming, one in community television and the other in pay-tv. Except for three community stations, all failed. In 1982 a third experiment started. Seven cable companies were granted licences for a period of two years to run a strictly regulated pay-tv service: no advertising, no exclusive coverage of sport or events, and no feature films less than one year old. The companies must subscribe, unlike the BBC and IBA, to the codes of the British Board of Film Censors and no 'X'-rated film can be shown before 10.00 p.m. Moreover, programme schedules have to be submitted in advance to the Home Office.

A national cable network will cost £2.5 billion, at 1982 prices, to bring cable to half the population – those living in the big towns. Capital will be raised by private investment but that will depend, say the investors and the entrepreneurs, on how far the government will relax its regulations of existing cable and pay-tv operations. People with other motives argue that, without strict regulation, cable television will follow the worst of American practices and undermine the quality of broadcasting in Britain, threatening the duopoly of the BBC and IBA.

A distinction has to be made between cable companies that operate as common carriers and those that act as publishers of programme material and services. The company which builds and owns the basic cable network could act as a commercial common carrier, operating like British Telcom and serving both television and business interests. Additional capital would have to be raised to install local networks in specific areas, hotels, blocks of flats or even whole towns. The companies which build and own these local networks could operate as common carriers, leasing channels to various television and business interests, each of whom would be his own publisher. Alternatively, they could themselves select the services to be relayed and be involved in the production of local programmes and the selling of advertising as in America. The company would then be a publisher, not a common carrier.

Whatever combination is worked out, and each area and each company could differ, a procedure would have to be established for the selection and licensing of individual cable companies. In America this is the responsibility of the local authority; in Britain, in so far as precedent has been established by local radio, it is the broadcasters, the BBC for public service stations and the IBA for commercial stations. Both work with the local authorities and have an advisory board representing local interests. The options open to government are to extend the responsibilities of the IBA and the BBC to local cable networks, alongside their local radio stations, or to create a new authority, or to do it themselves, as they do now with cable and the current pay-tv experiments. Persuasive arguments can be advanced for each option.

Government itself has first to make a decision about how it is going to deal with the communications explosion that will follow its executive acts. Will they see responsibility split between different departments, as now, or vested in a new single department to handle all aspects of television broadcasting, satellites, cable, pay-tv and home video, and video theatres. The inclusion of the film industry would make sense in that cinemas in the long term might become high-definition video theatres. There will also be a need for central co-ordination of, for example, education and religious broadcasting policy, if programme outlets proliferate. A department of this kind would also need to be party to the strategic planning of communications resources, to ensure that television and business interests are equally well served.

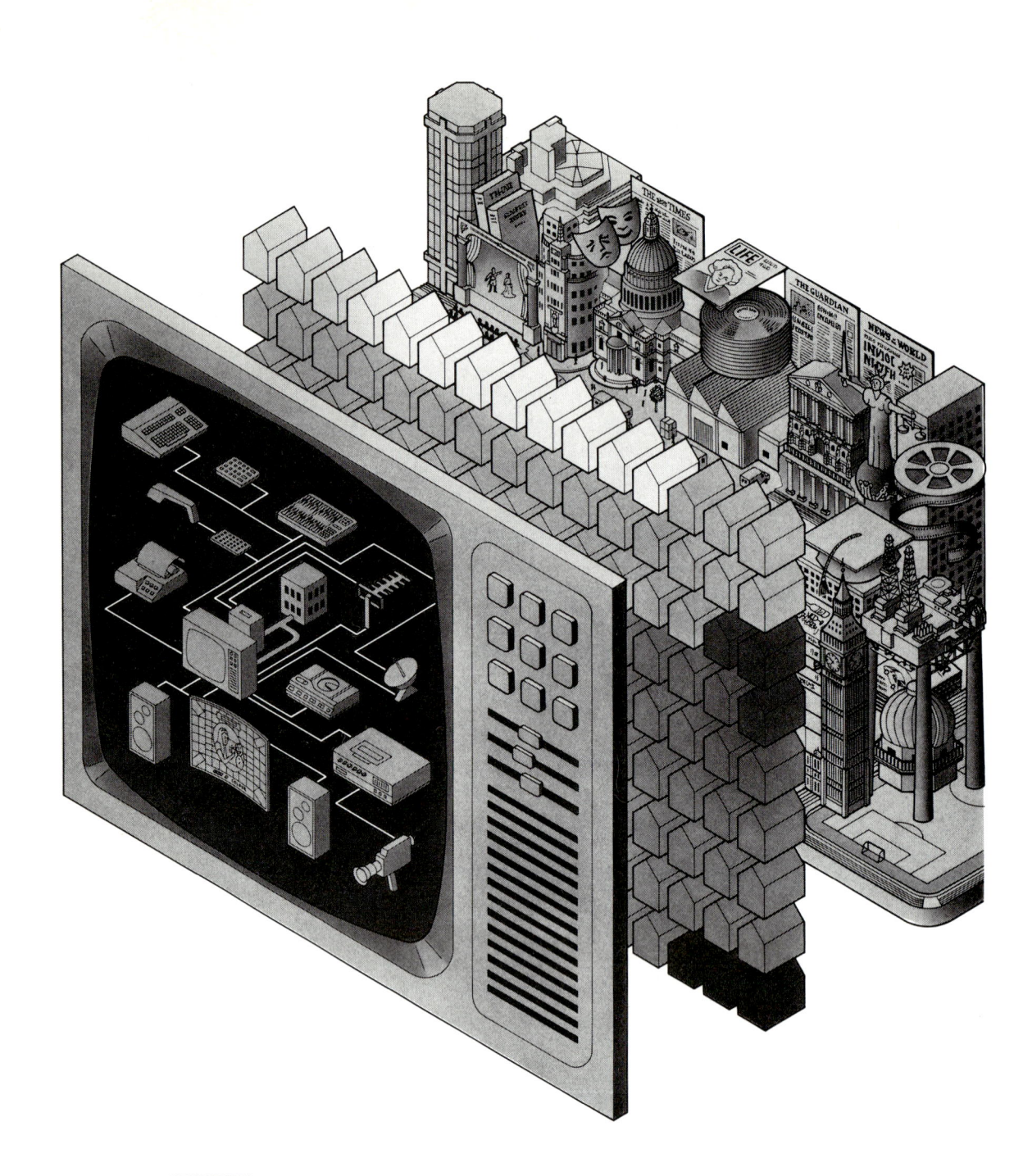

The elements of THE TOTAL EQUATION are the new systems and facilities that make up the home terminal; the increasing number of television homes throughout the world and the multiplicity of new sources of programmes and services.

11 THE TOTAL EQUATION

Television in the eighties in all its forms can be expressed as the elements of an equation, with the challenge to broadcasters on the one side and their response on the other.

The challenge comes from the growth of new systems of programme distribution and funding – DBS, cable, home video, video theatres, and pay-tv – and from the new sources of programmes and services – radio, records, cinema, theatre, sport, newspapers, magazines, books, education, religion, industry and commerce. The challenge will be manifest in new forms of competition for audiences, programmes, talent and money. The other side of the equation sets out the response of the broadcasters, diversifying into the new distribution systems to compete more effectively for audiences and for money, and seeking partnerships with the new sources of programmes and services, to remain competitive in the acquisition and provision of programmes, within technical and legal constraints and government regulation.

This total equation cannot be left to solve itself through market forces. That is the lesson from America. In countries with established broadcasting traditions governments will have to take steps to protect broadcasting if they wish to encourage the development of new systems. National broadcasting institutions, as major production houses, need to be kept intact. Each holds the critical mass of talent that ensures high programme and technical standards. Their transmission networks should also be kept in operation. A transmission network is permanent and secure. It is not easy to destroy, nor does it have to be replaced every seven years like satellites. It does not depend on privately owned resources like cable. Moreover, a transmission network usually has near total national coverage, and is 'public' in the sense that it is freely available to anyone with a television set.

Broadcasting should continue to be publicly funded by licence fees, by advertising or by a basic charge on the consumer similar to subscription television, or a combination of all three. It should not be by direct government grant, and broadcasting should remain free of direct government control. It should claim no special privileges over other markets, except that it should have statutory rights to certain topical events as part of its 'public' responsibility. Non-topical events could go first to one of the other markets, pay-tv, say, with a guarantee of a later showing on broadcast television. For competitive reasons, and to improve their services, broadcasters should be able to diversify into DBS, cable, pay-tv and video on a worldwide basis. Any profit from programme sales in these markets should be fed back into production, not to government. Broadcasters should be encouraged to foster partnerships to safeguard their programme sources.

DBS is an extension of television broadcasting, and in Britain it will be the responsibility of the BBC and the IBA by government licence. DBS should not be regarded as a substitute for television broadcasting. It is not secure and although it has a potential complete national coverage, reception in some areas will be critical.

Satellites are also an efficient means of distributing 'wholesale' to cable and television stations. In Europe they may be used for distributing national or pan-European superstations for relay by cable and television stations as well as for DBS. Satellites may also be used by minority and specialist groups, to link up their interests internationally. Satellite companies should be licensed by governments as common carriers, leasing their transponders on a commercial basis to a variety of users, each of whom, in law, becomes the 'publisher' of his own material. International agreement would have to be reached over possible abuses of satellite footprints spilling over into other national territory. Funding of satellite services could come from a combination of sources – advertising and sponsorship, donations, pay-tv, and rental charges from relaying countries.

Cable is efficient in urban areas in providing a large number of 'retail' outlets for broadcast television, satellite services, locally produced community and access programmes, and interactive two-way television. It can also provide a number of home information and security services as well as serving business interests. National Cable Authorities, created by but independent of government like the Broadcasting Authorities in the United Kingdom, should set technical standards, award licences for the construction and maintenance of cable networks for business and television use, define areas, and nominate the cable companies, possibly on a franchise basis as with commercial radio and television in Britain. Channels should be allocated to television stations and DBS services, to local programmes, and to home information and business services, under the terms of a licence or franchise agreement. Allocation of the remaining channels should be a matter for the cable company, though guidelines should be established by the Cable Authority. Funding of cable operations could come from advertising, sponsorship, pay-tv, and from rental charges on leasing channels. There could be public funding for local programmes.

Home video in the form of cassette recorders is well established as a means of recording programmes from broadcast television for playback at a later, convenient time. The practice will extend to all other sources of programming. There should be legal provision for this. Buying or renting video cassettes and video discs will be the only means of access to some programmes if the consumer cannot receive DBS signals or has no local cable service. Video programmes should be regarded as books or magazines and not be subject to any specific form of regulation.

Video theatres will exist in their own right or as an adjunct to other facilities such as a shopping precinct. They will show new feature films fed by satellites, pay-cable, or from video cassettes and discs, and will probably be first to exploit high-definition television. They will also feature televised 'live' events and foreign relays such as boxing matches. The programming should be decided totally by the management, and be funded by admission charges and advertising.

The new systems, DBS, cable, home video and video theatres, should be free to develop alternative programmes and services profitably, with minimum regulation. Equally, the quality of broadcast television should be sustained and minority and specialist interests safeguarded. Broadcasting in the sense that it is 'public' should be required to maintain a balanced schedule of programmes and services, which should be available to anyone with a television set. Such proposals would help balance the equation for television in the eighties and, hopefully, into the nineties.

INDEX

AAA (Amateur Athletics Association), 69
ABC (American Broadcasting Company), 14, 70, 71, 102, 104, 106
ABC Video Enterprises, 70
Acorn computers, 98
Action Committee – British Film Industry, 60
Actors' Guild of America, 113
ACTT (Association of Cinematograph, Television & Allied Technicians), 61
Advertising, 53, 106, 113, 120, 124
American Educational Television Network, 89
American football, 71
American space shuttle, 16
American Supreme Court, 59
Antiope – French videotex system, 27
Appalachian Community Network, 89
Aregon International, 98
Ariane rocket, 17
Art galleries, 67
Arts Council, 64, 65
ARTS Network, 58, 66, 70, 104
Associated Press, 76
Association of American Publishers, 82
Attenborough, David, 82

Baird, John Logie, 12
Barclays Bank, 96, 100
Basketball, 71
BBC (British Broadcasting Corporation), 14, 29, 41, 42, 46, 50, 67, 84, 85, 87, 92, 98, 103, 106, 108, 115, 121; charter, 117, 119, 120; diversification, 103; education, 87; income, 41; religion, 92; schedules, 103
BBC Enterprises Ltd, 103
Beatles, 55
Berne Copyright Convention, 113
Betamax – Sony video-cassette format, 22, 111
Big screens, 31
Blondie, 55
Books, 81–6
Bravo – American pay-tv network, 66

British Aerospace, 108
British Board of Film Censors, 120
British Home Stores, 96
British Petroleum, 96, 100
British Publishers Association, 82
British Rail, 100
British Telecom, 98, 100, 106, 121
Broadcasting, 12–14, 102–5, 123; audiences, 34; competition, 104–5; costs, 14, 34; diversification, 103, 104; funding, 106; growth, 37, 40; income, 38, 41; programme demand, 39, 42; regulation, 117; schedules, 43, 103
Broadway, 65

Cable, 18–20; audiences, 34; costs, 18, 20, 34, 121; developments, 20, 120; funding, 108, 121; growth, 18, 20, 37, 40; income, 39, 41; piracy, 115; programme demand, 39, 42; regulation, 117; schedules, 43, 107
Cable and Wireless, 100
Canadian Broadcasting Corporation, 21
Captain – Japanese videotex system, 27
Carfax – Radio information system, 51
Catholic Training Centre, Hatch End, 93
CATV (Community Antenna Television), 18
CBN (Christian Broadcasting Network), 94
CBS (Columbia Broadcasting System), 14, 25, 29, 31, 43, 58, 66, 71, 102, 104, 106, 109, 110, 112, 118, 119
Ceefax – BBC teletext, 14, 29, 104
Channel Four, 41, 77, 82, 84, 87, 103, 119
Chariots of Fire, 60
Chicago Tribune, 74
Cinema, 59–63
Closed-circuit television, 71
Coaxial cable, 110, 111
Collins, William, 84
Commanding Sea, The, 85
Computers – see Home computers

Confravision, 97
Cooke, Alastair, 82
Coppola, Francis Ford, 63
Coproduction, 85, 106
Copublishing, 85
Corporation of Public Broadcasting, 91
Cosmos (Carl Sagan), 82
Counterfeiting, 54
Covent Garden, 64, 65, 67, 72, 109
CRAC (Central Religious Advisory Committee), 92, 95
Cricket, 68
CTVC (Centre for Television and Radio Communications), 93
Cyclops, 87, 98, 99

DBS (Direct Broadcasting by Satellite) – see Satellites
Digital, 51, 112, 120
Discovision – MCA video-disc system, 97
Disney Productions, 48, 114

EBU (European Broadcasting Union), 17, 52, 111
Economist, 60, 78
Education, 86–92
Electronic Church, 93
Electronic Fund Transfer, 100
Electronic Mailbox, 98
Electronic Newspapers, 75
Electronic Printing, 76
Electronic Shopping, 99
EMI – see Thorn–EMI
English National Opera Company, 64
Entertainment Channel, 58, 66, 103, 107
Episcopal Network, 94
ESPN (Entertainment and Sports Programming Network), 69, 71, 104
Eternal Word Network, 94
European Court of Justice, 114

FCC (Federal Communications Commission), 29, 31, 100, 104, 107, 112, 117; fairness doctrine, 118; financial interest rule, 118, 119

126

Fibre-Optics, 19, 100, 110, 111, 120
Financial Times, The, 60, 73, 76
Footprints, 16
Ford Motors, 96, 124
Funding, 106–9

Gallup poll, 102
General Motors, 92, 96
Getty Oil, 69, 70
Goldcrest Films International, 60
Granite School District, 89

HDTV (High Definition Television), 31, 63, 112
Hearst, 70, 79, 104
Holiday Inn Inc., 97
Hollywood, 61, 66, 75, 82
Holography, 31
Home banking, 99
Home Box Office, 61, 66, 67, 70, 79
Home computers, 27, 53, 98
Home Office, The, 119, 120
Home taping, 54
Home terminal, 32–4
Hunt Committee, 120

IBA (Independent Broadcasting Authority), 111, 119, 121
IBM (International Business Machines), 97
IFPI (International Federation of Phonogram and Videogram Producers), 114
Indian cinema, 59
Industry and commerce, 96–101
IPC (International Publishing Corporation), 79
ITN (Independent Television News), 74
ITV (Independent Television), 41, 87, 92, 103, 106, 115

Japanese technology, 17, 37, 63, 96
JVC (Japanese Victor Company), 24

KCET, Los Angeles, 91
King's College carol service, 56
KWHY, Los Angeles, 74

Laservision – Philips video-disc system, 24, 25, 38, 84, 111
Legal issues, 113–16
Licence fees, 106
Lincoln Center, 66
L-SAT, 17

Maclean, Donald, 84
Madison Square Garden, 70
Magazines, 77–81
Magnetic video, 48
Market values, 36
Marks and Spencer, 96
MCA (Music Corporation of America), 25, 96
McCormack, Mark, 69
MDS (Multi-Distribution Systems), 14
Mercury, 100
Metromedia, 65
MGM Studios, 59
Miming, 55
Mitchell Beazley, 84
Mormons, 95
MTV Music Channel, 58, 59
Multi-screen cinema, 62
Museums, 67
Music, *see* Video Music
Music For Your Eyes Channel, 58

Nashville Music Channel, 58
National Jewish Network, 95
National Theatre, 64
National Westminster Bank, 96
National Youth Theatre, 65
NBC (National Broadcasting Corporation), 14, 102, 104, 106
Needletime, 54
News agencies, 74, 75
Newspapers, 72–7
Newsweek, 77
Newton-John, Olivia, 56
New York Herald Tribune, 18, 76
New York Times, 74, 76, 102
NHK – Japanese State Broadcasting Corporation, 31, 112
Nielson – American research company, 69
Nordsat, 17
Now!, 77
NTSC – American colour television standard, 12, 13, 25, 112

ONTV, Los Angeles, 70
Open University, 87, 98, 99
Oracle – IBA teletext, 14, 29
OTS Satellite, 111

PAL – Anglo–German colour television standard, 12, 25, 112
Pay-tv, 20–1; audiences, 34; costs, 21, 34; developments, 21, 120; growth, 38, 40, 41; income, 39; programme demand, 39, 42; regulation, 120; schedules, 45, 107
PBS (Public Broadcasting System), 14, 43, 66, 90, 106
Pearson–Longman, 60

Penthouse Channel, 79
Philips, 22, 24, 96, 111
Piracy, 54, 82, 114
Playboy Channel, 79
Playcable, 81
Pop-promos, 55
Prestel – British viewdata, 27, 98, 111
PTL – Religious Network, 94

QUBE – American interactive cable system, 19, 45, 84, 90, 100

Radio, 50–4, 87, 118
Radio-data, 52
Radio Télé-Luxembourg, 17
Radio Times, 79
Radiovision, 52, 53
Rank Organisation, 59, 62
RCA (Radio Corporation of America), 24, 98, 104
Records, 54–9
Regulation, 37, 117–21, 124
Religion, 92–5
Reuters, 74, 75, 76
RIAA (Record Industry Association of America), 114
RKO Pictures, 59
Rome Convention, 113
Rothschild, 108
Royal Opera House, *see* Covent Garden
Royal Shakespeare Company, 64
Royal Wedding, 103

Satellites, 14–18, 103, 110, 120, 124; audiences, 34; costs, 16, 34, 111; developments, 17, 63, 108, 111; growth, 38, 40; income, 39, 41; legal issues, 113, 116; programme demand, 39, 41
SECAM – French colour television system, 12, 25, 112
Selectavision – RCA video-disc player, 24, 38, 111
Sequential distribution, 48
Sesame Street, 90
Shellfax – Business communications, 96
Shopping guides, 99
Showcable, 103
Showtime, 66
SMPTE (Society of Motion Picture and Television Engineers), 111
Society for Promoting Christian Knowledge, 93
Society of West End Theatres, 64
Sony, 24, 31, 96, 114
South African Broadcasting Corporation, 69

Sponsorship, 64, 68, 106
Sport, 67–72
Stereo, 32, 55, 118, 120
STV (Subscription Television), 70, 107, 118
Sunday colour magazines, 77
Sunday Times, 74
Superstations, 45, 70

Tape levy, 54, 114
Technical development, 110–12
TED – early German video-disc player, 24
Teleconferencing, 97
Teletex – two-way electronic mail, 98
Teletext – one-way broadcast videotex, 29, 53, 81, 91, 110
Teletrack – closed-circuit television theatre, 71
Television standards, 12, 25, 111, 112
Telidon – Canadian videotex, 27
Telstar – first communications satellite, 14
Thames Television, 104
Theatre, 64–67
Thomson, 24
Thomsons, 85

Thorn–EMI, 56, 59, 84, 108
Time-Life, 77, 78, 85, 90
Trinity Broadcast Network, 94
Turner, Ted, 45
TV Guide, 102
TV Times, 79
Twentieth Century-Fox, 48, 60, 109

UHF (Ultra High Frequencies), 110
United Federation of Teachers, 90
Universal City Studios, 114
Universal Copyright Convention, 113
University of Arizona, 89

VHD – Japanese video-disc format, 24, 38, 111
VHF (Very High Frequencies), 51, 110
VHS – Japanese video-cassette format, 22, 111
Video albums, 55, 56
Video cassettes, 22–4, 124; audiences, 34; catalogues, 46, 55; competition, 34; costs, 23, 24, 34, 62; growth, 22, 38, 40; income, 39, 41; programme demand, 39, 42

Video Copyright Protection Society, 115
Video discs, 24–5, 110, 124; audiences, 34; catalogues, 46; costs, 24, 25, 34; growth, 38, 40; income, 39, 41; programme demand, 39, 42
Video formats, 22, 24
Video games, 26
Videograms, 22, 56, 113
Video music, 55–9
Video printer, 30
Video religion, 93
Video theatres, 62, 125
Videotex, 27
Viewdata, 27, 81, 118
Visionhire, 103

Wall Street Journal, 76, 102
Walt Disney, *see* Disney
WARC (World Administrative Radio Conference), 16
Warner Brothers, 84
WBLS – Music Channel, 58
Welsh Channel 4, 103, 119
Westminster Press, 73, 76
Wilson, Sir Harold, 60
Wimbledon, 67, 72